Amadeus William Grabau

Guide to Localities Illustrating the Geology, Marine Zoology
and Botany of the Vicinity of Boston

Amadeus William Grabau

Guide to Localities Illustrating the Geology, Marine Zoology and Botany of the Vicinity of Boston

ISBN/EAN: 9783337415457

Printed in Europe, USA, Canada, Australia, Japan

Cover: Foto ©berggeist007 / pixelio.de

More available books at **www.hansebooks.com**

A. A. A. S.

FIFTIETH ANNIVERSARY MEETING.

Boston, August, 1898.

GUIDE

TO

Localities Illustrating the Geology, Marine Zoology, and Botany

OF THE VICINITY OF BOSTON.

EDITED BY

A. W. GRABAU AND J. E. WOODMAN.

CONTRIBUTORS.

Prof. W. O. Crosby	. . . Geology; South Shore
" W. M. Davis .	. . . . . . Physiography
" B. K. Emerson	. . Geology; Turner's Falls Region
" W. G. Farlow	. . . . . . Marine Algæ
" J. E. Wolff .	. . . . . . Petrography
A. W. Grabau	. . . . . Marine Invertebrates
" "	. . . . Palæontology
J. E. Woodman	. . . . Geology; North Shore
" "	. . . Photographing and Collecting

A. A. A. S.
1898.

CONTENTS.

	PAGE
INTRODUCTION. J. E. WOODMAN	v

I.

PHYSIOGRAPHY. PROF. W. M. DAVIS.	1-7
Uplands of southern New England	1
Boston harbor	2
Provincetown, Cape Cod	3
Coastal plain of Maine	4
Connecticut valley: Meriden district	6
GEOLOGY: NORTH SHORE. J. E. WOODMAN	9-20
Plum Island dunes	9
Beach Bluff and Marblehead Neck	9
Nahant	10
Pine hill, Medford: "300-foot dike"	14
Gloucester moraine	15
Arlington moraine	15
Mystic river quarries, Somerville	16
Corey hill, Brighton	17
Newtonville esker and sand-plateau	18
Auburndale	19
GEOLOGY: SOUTH SHORE. PROF. W. O. CROSBY	21-31
The outer islands of Boston harbor	21
Nantasket and Cohasset	21-25
Nantasket beach to Green hill	21
The Nantasket ledges west of Hull street	23
The Cohasset shore	24
Northern Hingham	25
Mill cove, North Weymouth	26
The Monatiquot valley, Hayward creek (Paradoxides quarry), and Ruggles creek	27
Blue Hills of Milton	28
Quincy granite quarries	28
Lake Bouvé	29
Martha's Vineyard and Gay Head	30

	PAGE
GEOLOGY: TURNER'S FALLS REGION. TWO EXCURSIONS IN THE CONNECTICUT VALLEY. PROF. B. K. EMERSON	33–35
Surface deposits	34
PALEONTOLOGY. A. W. GRABAU	37–62
Nahant	38
Mill cove, North Weymouth	39
North Attleborough	39
Braintree quarry, Hayward creek	40
Bernardston	41
(Mansfield)	42
(Worcester)	42
Rockdale	42
Canton Junction	43
(Brockton)	43
Connecticut valley (Turner's Falls, etc.)	43
Gay Head	44–47
Indian hill	47
Highland light	48
Sankaty Head, Nantucket	49
Winthrop Great Head and other harbor drumlins	51
Gloucester	52
Boston	53–54
Brookline district	53
North bank of Charles river (Cambridge)	53
City point, South Boston	54
Literature	54–62
PETROGRAPHIC NOTES. PROF. J. E. WOLFF	63
PHOTOGRAPHING AND COLLECTING. J. E. WOODMAN	65–66

II.

ZOÖLOGY: MARINE INVERTEBRATES. A. W. GRABAU	67–96
Revere beach	67
Swampscott beach	76
Nahant neck	77
Beverly	78
Castle rocks, Nahant	85
East point, Nahant	87

III.

BOTANY: MARINE ALGÆ. PROF. W. G. FARLOW	97–100

INTRODUCTION.

It is the aim of this guide to reach especially members of the Association present at the Boston meeting, who may desire to visit localities not included in the general excursions offered by the hosts of the occasion. It is not intended, therefore, to be complete for the region; but gives as it were sample cases which are likely to be of interest to visitors. The literature appended, with one exception, contains only papers which are recent and easily accessible. In the case of Palæontology, however, the list is probably fairly complete. Geologists from the interior cannot fail to be struck with the marine action exhibited; and those acquainted mainly with fossiliferous rocks of simple structure will appreciate the change to a district composed largely of igneous types, with sediments of complex history. Biologists also will appreciate the opportunity for studying the local marine fauna and flora in their various phases. Most of the localities listed are within a short distance of Boston, and may be visited in a portion of a day. A few others have been noted, because their interest entitles them to recognition, and because some may desire to see them.

To all those who have aided in making the guide we offer our thanks, appreciating the fact that the time has been short in which to write the notes presented here, and that it had to be taken from other important duties. In some instances, small fragments of material have been contributed by others than the authors of the articles, and these are acknowledged in each case. The illustrations have been selected, not so much to illustrate the text as to call additional attention to the localities; and we thank the owners of the plates for their kindness in allowing their use. Each illustration is credited to its source.

It was intended originally to insert in the paper a section upon land plants; but it has been deemed unnecessary, because the New England Botanical club has made arrangements to conduct excursions and to furnish guides for visitors.

J. EDMUND WOODMAN.

Geological Laboratories, Mus. Comp. Zoöl.,
   Cambridge, Mass.
      August, 1898.

I.
PHYSIOGRAPHY.

Prof. W. M. Davis.

UPLANDS OF SOUTHERN NEW ENGLAND.

The view from any of the hills near Boston, preferably from Arlington Heights (reached by electric cars, passing through Cambridge), discloses the moderate relief of the skyline, little in accord with the great disorder of the rock structures. This has given rise to the opinion that the skyline represents the general level to which denudation reduced the deformed structures, when the whole region stood lower than now; and that the valleys by which the uplands now are so freely dissected result from a later cycle of denudation, which was introduced by the uplift of the region to an altitude somewhat greater than that of to-day. A slight depression after the valleys had been eroded was the chief cause of the existing irregularity of the shore line, subject to modification by slight oscillations of level, and by plentiful deposits of drift in connection with the glacial period.

In the neighborhood of Boston, the area occupied by rocks of moderate resistance is so great that an extensive lowland has been worn down, known as the "Boston Basin" in its topographical (not strictly in its geological) sense. Here the overlapping sea enters farthest into the coast line by reason of the lowland, and this has given Boston an advantage over the neighboring early settlements of Plymouth and Salem. The harbor would enter still farther into the land, but for the drift that floors much of the basin.

The uplands in eastern Massachusetts are so extensively interrupted by valleys that it would be difficult to convince anyone of the reality and continuity of the ancient peneplain which the uplands are thought to represent. An excursion farther into the interior is needed to emphasize the contrast between the comparatively even and extensive uplands and the relatively narrow and steep-sided valleys that interrupt them. Gardner, on the Fitchburg railroad, is a good point to illustrate these features. A drumlin crowned by a reservoir just north of the village offers an extensive view, including Wachusett in the southeast and Monadnock in the northwest. A walk over Hoosac mountain in western Massachusetts is a still more instructive excursion in this regard: Deerfield valley, 1,000 feet or more deep, being almost cañon-like in contrast to the Berkshire highlands in which it is incised. The view from West Peak, near Meriden, Conn., is perhaps more satisfactory than any other. By considering all of these features together, it is believed that the explanation of the uplands and valleys of southern New England in the manner above suggested gives a reasonable account of them; but it should be understood that the peneplain of the uplands was never very smooth, and that in many parts of the district it is dissected so thoroughly to-day as to be recognizable with difficulty.

Literature.

Davis, W. M. — The geological dates of origin of certain topographic forms on the Atlantic slope of the United States. (Geol. Soc. Am., Bull., vol. 2, pp. 541–542, 545–586.)

Davis, W. M. — The physical geography of southern New England. (Nat. Geog. Monographs, vol. 1, No. 9.)

Davis, W. M. — Geographical illustrations. Harvard University, 1893.

BOSTON HARBOR.

The most notable features of Boston harbor are the drumlins which stand forth as numerous islands, and the extensive beaches which tie many of the islands to each other and to the mainland. The bed-rock borders of the coastal reëntrant are found at Lynn and Nahant on the north, and at Nantasket and Cohasset on the south; but the harbor proper is much restricted from these limits, by the group of drumlins extending from East Boston to Winthrop. An afternoon excursion by boat to Pemberton landing (Hull), thence by rail to Nantasket and back to Boston by boat, forms a pleasant and instructive outing. On the way down the harbor, many drumlins are passed in various stages of destruction. One of the outermost islands, bearing "Boston light," exhibits the remains of a very large drumlin, from which more than half the original volume has been swept away on the exposed eastern side, and about a sixth on the less exposed western side. A long double-curved spit, bare at low tide, trails into the harbor from this island. It will be noticed that the vigor of attack of the sea at various points is shown by grassed and bare erosion slopes. In a few cases, however, the turf is laid artificially over fortifications. Pemberton landing is on the western end of a shorter spit formed by waste from the drumlins of Hull. The railroad thence follows the shore for a mile eastward. Leaving the train at Allerton, the axis of a long drumlin may be followed eastward to the fine bluff of Point Allerton; thence a walk of a mile southward along the beach leads to Strawberry hill (bearing a reservoir), once cliffed by the sea, but now protected by several hundred feet of beach built in front of it. Faint lines of former beaches may be traced north and south from the abandoned sea cliffs of this hill. (Take train from Waveland station to Nantasket.) Other abandoned cliffs are found on drumlins just north of Nantasket, beyond which the ragged rocky coast extends southeast towards Cohasset, with the ledges of Minot's light off shore. The irregularity of the rocky shore line suggests that little detritus has been brought to Nantasket beach from the southeast; and hence that the materials of the beach have been supplied largely from destroyed drumlins, whose bases now remain in the shoals marked by beacons east of the beach. (See also Geology, p. 21; Palæontology, p. 51.)

Literature.

See map in Crosby's Geology of the Boston Basin, Part I, Nantasket and Cohasset (obtainable at Museum of Bos. Soc. Nat. Hist.).

PROVINCETOWN, CAPE COD.

"The bended arm of Massachusetts," as Thoreau called Cape Cod, consists of (Tertiary?) bedded sands and clays on which a greater or less amount of glacial drift has been deposited, sometimes in strong morainic form, sometimes as broad washes of gravel and sand. The outline of the forearm of the cape when the present attitude of the land was assumed has been modified greatly by wave action. A long, smooth, slightly convex cliff has been cut fronting the ocean and along the eastern side of the cape, and extensive sand reefs and spits have been built northward to the

Provincelands and southward beyond Chatham to Monomoy. A two-day trip from Boston, across Massachusetts bay and back by steamer, with two half-days about Provincetown, will repay richly any geologist from the interior, to whom the features of a strongly-worked shore line are a novelty. By driving at once to Highland light, on the arrival of the boat in Provincetown, several afternoon hours may be enjoyed on the bold bluff and the superb beach below it. Facetted or wind-carved pebbles may be gathered in abundance just north of the signal station on the bluff. While returning to Provincetown, a walk over the upland leads to High Head, the northernmost point of the mainland of the cape, whence the long curved spit of the Provincelands stretches to the northwest, bearing extensive sand dunes, old and new, tree-covered and bare. From the same point a good view is gained of the abandoned sea cliffs of High Head, now enclosed by the sand reefs and marsh of the Provincelands. Near by, the cliffs cut by the waves of Cape Cod bay make an obtuse angle with the abandoned cliffs. The following morning, a pleasant walk leads out to the great dunes that are invading the forest northwest of the village. The action of the wind in building may be studied here to great advantage. (See also Palæontology, p. 48.)

Literature.

Davis, W. M.—Facetted pebbles on Cape Cod, Mass. (Bos. Soc. Nat. Hist., Proc., vol. 26, pp. 166-175.)

Davis, W. M.—The outline of Cape Cod. (Am. Acad. Arts and Sci., Proc., vol. 31, pp. 303-323.)

Grabau, A. W.—The sand plains of Truro, Wellfleet and Eastham. (Science, new ser., vol. 5, pp. 334-335.)

Upham, W.—The formation of Cape Cod. (Am. Nat., vol. 13, pp. 489-502, 552-565.)

COASTAL PLAIN OF MAINE.

The composite topography of the coastal district of Maine is shown well from the summit of Blackstrap hill, a large drumlin six or eight miles northwest of Portland. The general evenness of the skyline can be seen to the east and west; a number of monadnock-like mountains rise above it in the north, Mount Washington being one of the farther summits. The uplands, generally composed of crystalline schists, are dissected thoroughly into rocky ridges, separated by wide-open valleys that have been worn to a depth of several hundred feet. It is believed that these valleys

EXPLANATION OF PLATE.

From the article "Cape Cod" by W. M. Davis, in Nat. Geog. Mag., vol. 9.

Fig. 1. Stages in the development of the outline of a barrier reef at Cape Cod.

Fig. 2. Ideal section showing development of shore profile on Cape Cod.

Fig. 3. Map of stages in the gradation of an irregular shore line, that which Cape Cod formerly possessed.

EXPLANATION OF PLATES.

From the "Outline of Cape Cod," by W. M. Davis, in Am. Acad. Arts and Sci., Proc., vol. 31.

FIG. 1. Stages in the development of the outline of the north end of Cape Cod.

FIG. 2. Ideal section showing development of shore profiles on Cape Cod.

FIG. 3. Map of stages in the gradation of an irregular shore-line, like that which Cape Cod formerly possessed.

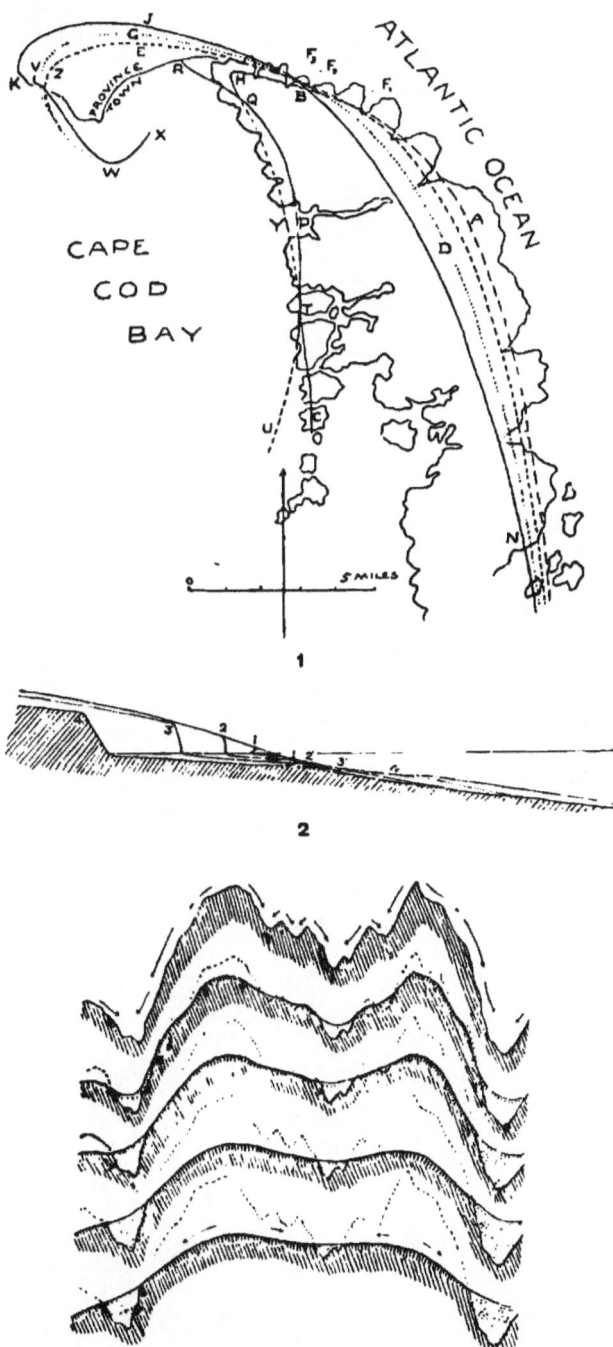

EXPLANATION OF PLATES CXIII–III

Fig. 1. Sections illustrating the growth and development of Alum style.
Fig. 2. Map showing the progress of erosion and subsequent formation of Alum style.
Fig. 3. Map of the change in the outline of Hilo Head shown.

EXPLANATION OF PLATES—CONTINUED.

FIG. 4. Sections illustrating the growth and subsequent history of offshore bars.

FIG. 5. Map showing the progress of erosion of headlands and the migration of flying spits.

FIG. 6. Map of the changes in the outline of High Head and vicinity.

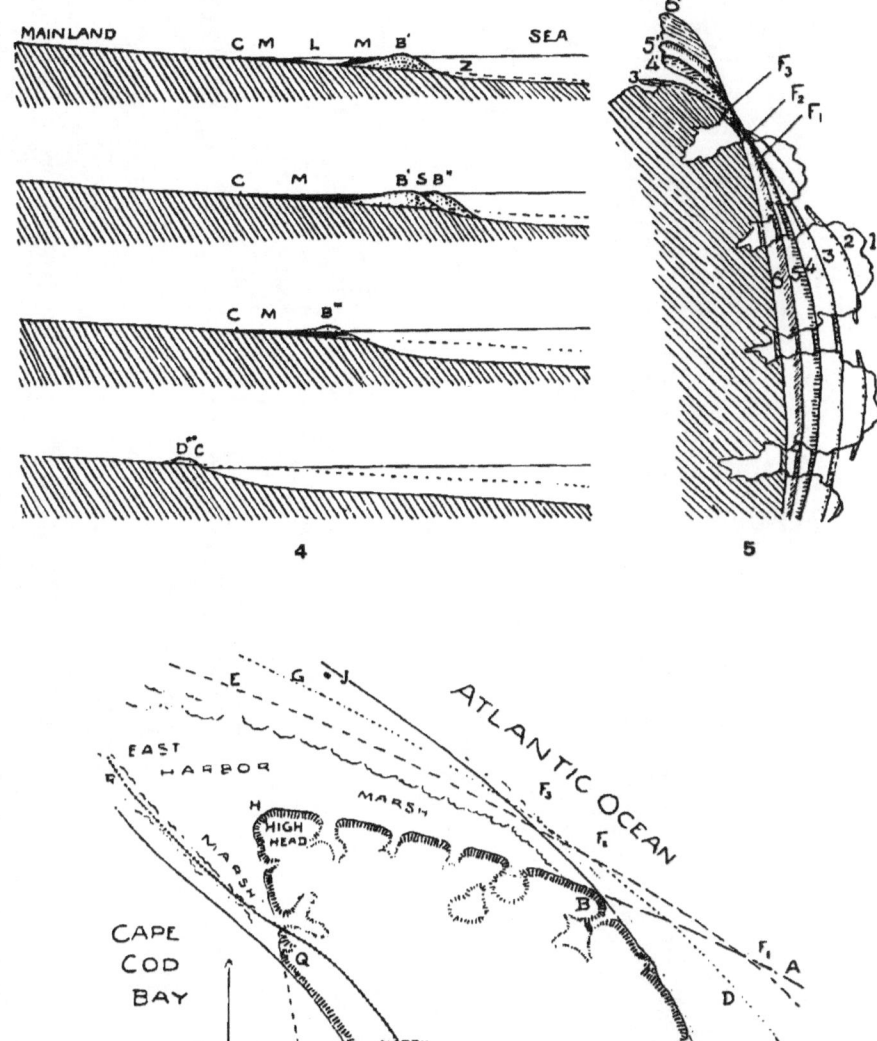

represent the Tertiary dissection of the uplifted Cretaceous peneplain, whose upland remnants constitute so important a share of the topography of the Atlantic slope.

Submergence of the valleys produced the ragged coast of Maine, the depression being associated with the glacial epoch; but since then there has been an emergence of at least 300 feet, as may be proved readily on this excursion. During the period of subsidence, postglacial marine clays were deposited in the arms of the sea and are to be seen now in the valleys, where they form flat fields dissected by a young drainage system, with narrow valleys and frequent landslides. The shore line of the 300-foot level may be recognized in a bench and bluff at the southeast base of Blackstrap hill, and again at Poplar hill, another drumlin three miles farther north, as well as on certain beach-like gravel flats, north and south of the latter. The present ragged shore line of Maine is therefore not the direct result of the submergence of a rugged land, but of the emergence of an uneven sea bottom—uneven because the marine clays that were spread upon it had not been deposited in sufficient quantity to smooth over its previous inequality.

The farmers of the coastal district make a division of their land on a strictly physiographic basis. The ledgy ridges are left to forest, wood lots, and rough pastures. Ridges of this kind advance between the clay-filled valleys toward or to the coast line; outlying ridges or hills form the island fringe off shore. Till-covered uplands are generally cleared and farmed, as north of Blackstrap hill; here stone walls often divide the fields. The clay-filled valleys are cleared and cultivated; the roads are very bad in wet weather, unless improved with gravel. Farmhouses are located frequently close to the line between ledgy hills and clay fields. Often streams are superposed locally on ledges once buried by the clays; hence waterfalls are common near the shore line, and this feature gives reason for the occurrence of paired cities, like Lewiston and Auburn on the Androscoggin, and Saco and Biddeford on the Saco.

A comfortable trip from Boston to Portland and back may be made by boat at night; and the day between will suffice to give a good view of the coastal district, especially if a bicycle is taken along, and the roads are not rough and muddy from recent rain.

Literature.

There is no literature upon this subject, from the standpoint presented here.

CONNECTICUT VALLEY: MERIDEN DISTRICT.

A single day's visit to the immediate neighborhood of Meriden, Conn., will give a good view of the general features of the Triassic formation, between the crystalline uplands on the east and west. Reaching Meriden by evening train from Boston, the following early morning may be given to the ascent of West Peak, the highest of the Hanging hills, about three miles northwest of the city. These hills are formed on the main extrusive trap sheet of the valley. The summit commands a fine view of the western uplands and valley lowlands; the former being a peneplain of Jura-Cretaceous denudation, now uplifted and dissected by relatively narrow valleys; the latter being a rough local peneplain, the product of denudation in some part of Tertiary time, surmounted by residual trap ridges, ornamented with drumlins, and veneered with washed drift. In clear weather, Long island is seen distinctly beyond the Sound.

Lane's quarry in the main trap sheet, a mile north of the city, may be visited next. It exposes the upper vesicular surface of one lava flow, buried under the dense basal portion of a second flow; the compound mass being faulted. The fragmental deposits of the anterior lava sheet, locally known as the Ash Bed, are exposed about three miles northeast of Meriden on the road to Berlin. An active walker might cross the fields from the last point and ascend Chauncey Peak, from whose southern bluff an excellent view may be obtained of the several blocks into which the district is divided by faults. A characteristic contact of the overlying sandstones with the vesicular upper surface of the main trap sheet is found in Spruce brook, at the northeast end of Lamentation mountain, a few hundred feet south of a cross-road and about five miles from Meriden; but this extension of the trip would require a horse and carriage.

By the use of a team, a second day in this locality might include a visit to the basal contact of the Triassic sandstones and conglomerates on their crystalline foundation, displayed well in the ravine of Roaring brook, three miles west of Southington; and the upper contact of the overlying sandstones and shales with the intrusive trap of Gaylord's mountain (the northern extension of the West Rock ridge series) in another Roaring brook, three miles southwest of Cheshire. A third day would allow an excursion to Middletown and the great sandstone quarries of Portland, where the Connecti-

EXPLANATION OF PLATE.

From "Quarries in the lava beds at Meriden, Conn.," by W. M Davis, in Am. Journ. Sci., 4th series, vol I.

FIG. 1   Map of the general relations of trap and faults in the Meriden district. The Meriden quarries are located at the point marked Q.

FIG. 2   Diagrammatic map and section of the local relations of the quarry ridge at Meriden.

FIG. 3   Diagrammatic map and section showing the location of the quarries and surrounding country at Meriden.

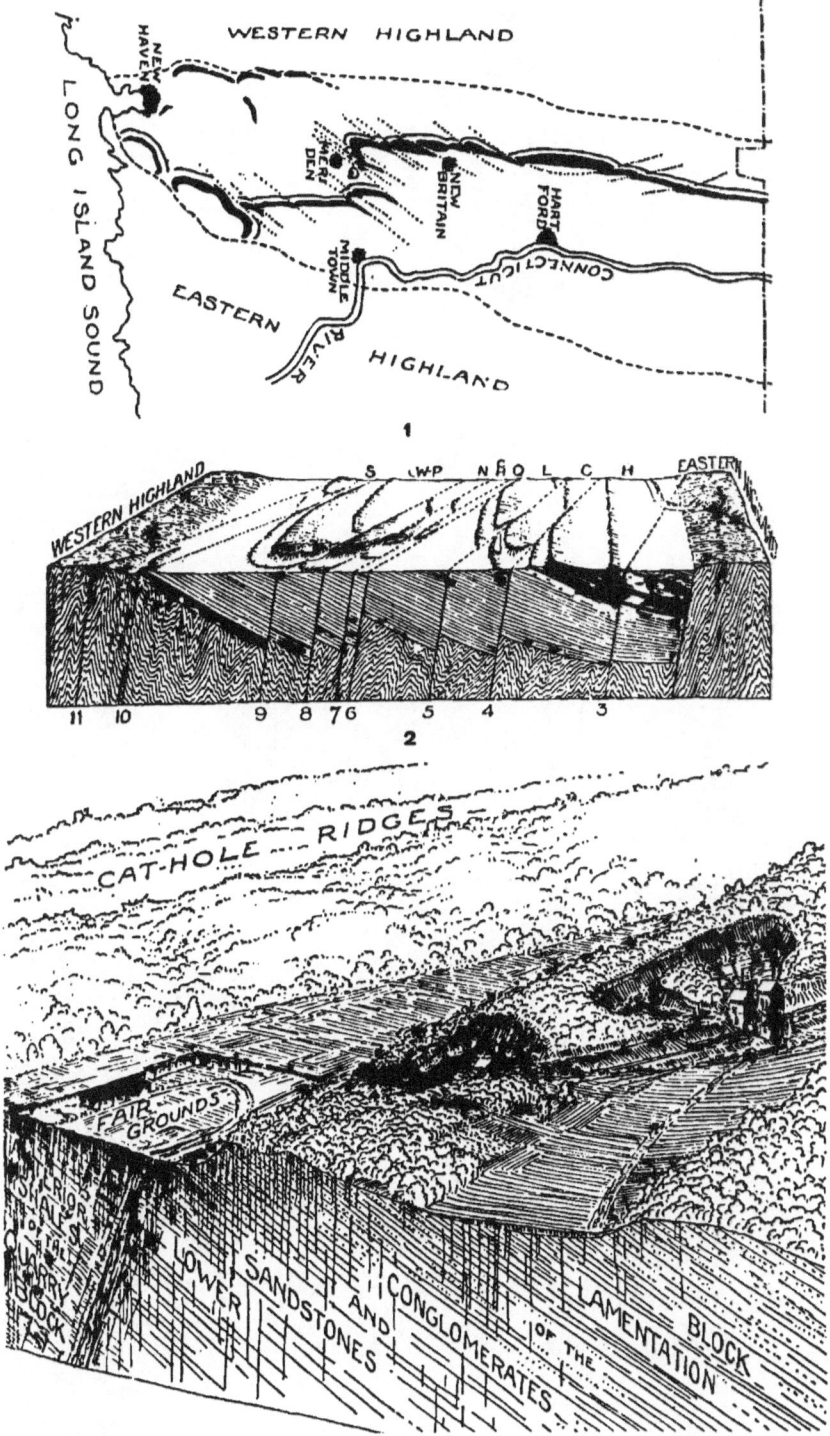

cut leaves the broad Triassic lowland and enters the narrow gorge-like valley that it has cut in the eastern uplands on the way to the sea at Saybrook. A beautiful view of this district may be obtained from Great hill north of Cobalt station, Air Line railroad, a few miles east from Middletown. As an illustration of a narrow valley worn in uplands of resistant rocks, draining a broad valley lowland underlain by weak rocks, this view has few equals.

Literature.

The literature of the Connecticut valley is abundant, but usually views the subject from the geologic standpoint. Among physiographic papers is the following:—

Davis, W. M.—Topographic development of the Triassic formation of the Connecticut valley. (Am. Journ. Sci., 3d series, vol. 37, pp. 423-434.)

Of geologic papers, there are many by Dana, Davis, Emerson and others, including the final report upon the Connecticut Trias, by Davis, which is in press as this is written.

Routes.—The routes to the main points mentioned are as follows:—

Boston harbor, any of the excursion boats plying between the city and resorts on the south shore.

Provincetown, by rail, Old Colony system from Kneeland street station. By boat, each morning from Commercial wharf, Boston; excursion stop-over tickets, $1.50.

Coastal plain of Maine, by train, Eastern division, Boston and Maine road from North Union station, for Portland. By boat, from India wharf, Boston, for Portland; single fare $1.00.

Meriden, Boston and Albany railroad to Springfield, Mass.; New York, New Haven and Hartford road thence to Meriden.

GEOLOGY: NORTH SHORE.

J. Edmund Woodman.

PLUM ISLAND DUNES.

Route. — (1) From North Union station by Eastern division, Boston and Maine railroad, to Newburyport; electrics eastward from station to near Merrimac square; horse cars from Merrimac square to Plum Island, north end. Single fare $1.00. (2) By same rail route to Ipswich; walk to boat; take boat for Plum island, south end.

Fine dune action, constructive and destructive, can be seen here — in operation if there be a wind. The formation of eolian ripples, structure of dune-sections, encroachment of wind-blown sand westward, swamping up of protected shallow-water areas between the mainland and the off-shore bar forming the island, protection by beach grass, and filling of lagoons by eolian sand, are among the phenomena visible. Marine action itself is not so varied in its effects as elsewhere.

BEACH BLUFF AND MARBLEHEAD NECK.

Route. — By rail, Eastern division Boston and Maine road, Marblehead branch, from North Union station, tickets for Beach Bluff; fare $.30.

By wheel, to Lynn by way of Broadway, Chelsea; thence along line of Swampscott electrics to within one block of Beach Bluff station. Turn to right in either case, to water front.

Marine action is powerful all along this part of the coast. A pretty continuous bench can be followed north to the wall beach and in places elsewhere. Between Marblehead Neck and the mainland to the south is a magnificent boulder beach, on which all the phenomena characteristic of such a form occur. The rocks composing it represent all the phases to be found between Marblehead and Swampscott. A great variety of erosion forms occur in the rocky portion of the coast, due to primary or secondary structures, dikes, etc. The protective effects of barnacles and seaweeds can be studied successfully.

On the east side of the shore road, a few hundred yards north of Clifton station, is a striated ledge of granite, with its lee side toward the east. This side presents the typical curves of marine erosion, accomplished apparently after most of the glaciation. The whole was buried in stratified drift, and has been exposed only recently through road building. The surface of the ledge is about fifty feet above the sea.

The region possesses great petrographic interest. From about the middle of Swampscott northeast the rocks have the same general characters almost to Marblehead Neck. The odest is diorite, into which at an early stage a granitite has intruded. The relations between the two are various. In general, the contacts are dim and intricate. At one place the older rock prevails, at another the younger; and in many ledges, the mass is a breccia of granitite carrying irregular horses of diorite. Into these rocks a lighter granitite and an eleolite syenite have penetrated, and lastly a black trap. This is largely in dikes which run roughly parallel to the coast, and have well defined walls, often separated from the country rock. Since the last intrusions, which are probably Mesozoic, a great amount of faulting has occurred, much of it in a southeast direction. The "Lincoln dike," off Clifton, is the largest and best-known case. In the trap basaltic structure is roughly developed in some cases, and many dikes have porphyritic centers.

Farther north, including the southern part of Marblehead Neck, the main rock is porphyrite. Beyond here it is a coarse granitite for the most part, on the ocean side. The well-known Bostonite occurs on the west side of the neck, but can be found only at low tide, and by diligent search.

Literature.

Sears, J. H. — Various articles on the geology of Essex county, in Essex Inst. Bull.

NAHANT.

Route. — *By rail*, Eastern division Boston and Maine road, from North Union station, to Lynn; fare, $.20.

By electrics, Lynn and Boston road, Scollay square to Lynn; round trip fare, $.25.

By wheel, over Charlestown bridge, through City square, Charlestown; over elevated bridge to Chelsea, through Broadway, Chelsea, to Lynn. From Lynn, by barge, wheel, electric, or afoot, to Lynn beach, walking thence southeast to Eastern point on Great Nahant, making observations *en route*. Return to Lynn may be all the way by barge if desired, for most of the trains; barge fare, single, $.15, round trip, $.25.

Lynn beach, from Lynn to Little Nahant, offers an excellent opportunity for the study of marine constructive action. The bouldery character of Marblehead Neck is absent except at the ends. The variations in texture along its course and upward from tide-line to crest can be seen plainly. The remnants of a few dunes stand by the roadside. Ripple-marks, wave-marks, rill-marks, trails, impressions of organic forms, etc., are abundant at low tide. Rafting in of pebbles and shells attached to *Laminaria* is a common sight, especially after easterly storms. The spit-building has progressed from both ends; but there is such a preponderance of southward motion, that, as can be seen in the field, the union of the two spits was close to Little Nahant. The beach is very recent, and growing rapidly. (See also Zoölogy, p. 77; Botany, p. 97.)

Little Nahant consists of syenite, and a small amount of sedimentary rock on its northwest side; the latter quartzite and metamorphosed green slates, with a strike N. 50° E., and a dip 70° S. E. The whole is intruded by trap dikes of diabase or basalt, probably of the same late series as those at Clifton and in Somerville. The glaciation offers little of interest. In what appear to be sheltered spots on the northwest slope, some eolian action is in progress. Marine effects are shown here and at Great Nahant as well as anywhere on the coast, especially with reference to lines of weakness. The cliffs of Little Nahant give indication on the north side of a twenty-five foot marine bench, possibly postglacial. The appearance is clearer on the east end of the land. Numerous dike chasms, especially on the south side, give effective exhibitions of the wearing action of the waves at high tide, by their ceaseless pounding and reflow. Joint chasms present the same appearance; and in one instance—"Irene's grotto," on the southeast side of the peninsula— a large section of rock has disappeared below, leaving a solid roof above. The place hardly can be dignified by the name of cave, but is only an arch ten or fifteen feet long. Protective effects of algæ and barnacles are displayed to great advantage on the broad rock shelf which extends from the eastern end at tide level.

Between Little and Great Nahant is a fine wall beach, high and coarse at each end. The gradation from angular blocks to well-worn pebbles can be followed in all its stages. At the north end is a partial dead-water, wherein great masses of kelp lodge. On the back side of the beach many angular pieces may be found, thrown up chiefly during the storms of winter. At times in this season the road is impassable. It is probable that the "pocketing" of the material is due, not to currents from either end, but to the greater blow and transporting power of long waves on the two horns, and the rapid loss of this power as the wave moves toward the apex of the angle.

Great Nahant is composed of two islands, and the lines of union which have made them one can be seen clearly. The swampy area running south between the main hills and Bear pond marks the position of the old strait. The main igneous rock is a coarse diabase. On the south and southeast are lower Cambrian sediments. They consist chiefly of indurated pelite and white limestone; the former containing calcareous concretions, which are represented largely by cavities filled with epidote crystals. They are more strongly marked below the sheets of diabase, and become less frequent farther away.

The limestone is fossiliferous, bearing indistinct outlines chiefly of *Hyolithes communis* var. *emmonsi*. The strike of the series is N. 50° E., its dip about 45° N. W. Trap dikes intersect both sediments and diabase. Just beyond the outcrop of the white limestone beds a transverse fault may be seen. The beds on the north have been thrown down thirty or more feet, bringing the white limestone near sea level. The main trap sheet, which on the south of the fault is exposed in the field south of the stone building, north of the fault outcrops along the cliff walk. Below it is the best place to observe the altered calcareous concretions. The fault line is marked by a fine chasm east of the cliff walk.

The difference in effect of the marine action upon the igneous rocks and clastics is brought out clearly here. The former have the rounded outlines seen elsewhere on the coast. The latter are angular, bold, rugged. Several systems of joints intersect the strata, which dip landward; hence the irregularity of the projections, the presence of stacks, joint chasms of great length, etc. Dike chasms are abundant. The best is the "Devil's bridge," a little north of Pulpit rock (the best-formed stack). Here a part of the dike remains and forms a natural bridge. The action of the waves at high tide is worthy of close attention.

On the north side of Great Nahant, a fair bench has been formed by the sea. Follow the cliff walk southward, crossing a coarse pebble beach to the next rocky headland, to where the north-south road ends. From here, in front of a dwelling house, a path leads down to Swallow's cave, perhaps the finest cave on the coast. It can be explored only at low tide, and will be found to extend through the cliff from end to end, and to terminate upward in an arched roof, broken only by one circular window. Near the outer end the tunnel is divided by a projecting piece of the country rock, where the eroded dike itself formerly was divided. The tunnel is long and high and narrow, and on its bottom are several tidal basins filled with water. For this portion of the trip, arrangements can be made with the barge driver to call at the cave at any hour, thus saving further walking. The spouting horn, a miniature dike cave, can be visited by leaving the barge on the north shore (the main road to East point) just east of Castle rocks, where a path leads northward to the horn, whose location is marked by a bench on the summit of the cliff. The visit should be made at half-tide, and the view gained from below. Here, as so often on this coast, care is necessary to prevent a repetition of the too frequent accidents. A satisfactory trip can be made by asking the barge driver to stop first at the spouting horn at the ebb tide. He will drive to a point west of the horn, whence a path leads eastward. Follow the path east and south beyond the horn, striking the road near Castle rocks. A short walk will bring one to the entrance to the Lodge estate. Walk from here by the foot-path to East point, going later to the cave as already directed. (See also Palæontology, p. 38; Zoölogy, pp. 85, 87; Botany, p. 97.)

Literature.

Foerste, A. F. — The paleontological horizon of the limestone at Nahant, Mass. (Bos. Soc. Nat. Hist., Proc., vol. 24, pp. 261-263.)

Lane, A. C. — The geology of Nahant. (Bos. Soc. Nat. Hist., Proc., vol. 24, pp. 91-95.)

PINE HILL, MEDFORD.

Route.— *By rail*, Boston and Maine road from North Union station to Medford, five miles. Single fare $.10. By electrics, Medford and Malden cars from Scollay square to Medford square.

By wheel, cross Charlestown bridge east of North Union station, through City square and Sullivan square, Charlestown; Winter Hill avenue and Mystic avenue to Medford square.

Continue through Medford square by all routes, north for one-half mile to point where sign " Middlesex Fells Reservation " appears on left. Turn off along driveway, keeping to right at junction of roads. Dike appears here underfoot and for nearly a mile north.

A dike of diabase, commonly known as the " 300-foot dike," lies in the main between two hills of country rock. Along the road, diabase soil can be contrasted with glacial soil. The age of the erosion can be noted, and the contact on either side followed quite closely. On the west, at the south or nearer end, it is very involved; and at various places along the margin on both sides other dikes having an east-west strike can be found. The country rock in the southern portion of the field is porphyrite; in the northern, granitite. The relations of the two are not shown. The former varies considerably in the size and abundance of the orthoclase crystals. The diabase also varies much in texture from center to sides, becoming in some places a fine-grained trap. It is probable that parts show the microstructure of basalt rather than of diabase. At the north end of the walk, where the dike is lost to sight for some distance, a small section of the east contact indicates a slight dip to the westward away from the vertical. What is probably the same dike appears also at the old Powder house in Somerville, and in a quarry on Granite street in the same city. The former shows inclusions of rocks which lie beneath the sediments of the Boston Basin.

Perhaps the most interesting feature of the dike is the weathering. The southernmost quarry exhibits the concentric arrangement well, and all stages from compact rock to fine gravel can be seen. In any case, however, slight changes have taken place, and the powdered rock effervesces freely with hydrochloric acid. To the south near Medford square, along Governor's avenue, road-cuts also show residual boulders and concentric peeling. Good photographs can be obtained.

Literature.

Jagger, T. A., Jr.— An occurrence of acid pegmatite in diabase. (Am. Geol., vol. 21, pp. 203-213.)

Merrill, G. P.— Rocks, rock-weathering and soils. 1897, pp. 218-222.

GLOUCESTER MORAINE.

Route.—By boat, from Central wharf, Boston, to Gloucester; single fare, $.50 ; round trip, $.75.

By rail, Eastern division Boston and Maine road, to Gloucester; fare $.72.

From Gloucester, take electrics for Rockport, alighting when most convenient; or better, walk eastward along electric track, turning south at any desired point. The excessively rocky portion lies chiefly on that side. Again, take electrics for Lanesville, alighting at Riverdale, and walking up to the morainic field. Visit Rail Cut hill. A good view can be obtained from near the signal staff, and a heavy moraine lies west of the hill.

All the central part of the island of Cape Ann is occupied by morainal material; but the portions most interesting are the "dogtown commons," formed of excessively bouldery drift, and so stony as to be worthless for cultivation. These form some of the roughly parallel ridges running northeast. They lie mainly in two or three lines, on both sides of the railroad to Rockport. The electric line is nearer the outermost or southern ones. Their position can be found best from the map opposite page 608 in the paper noted below. Among other interesting problems connected with the region are differential postglacial weathering, postglacial and preglacial stream erosion, and the composition of the moraine.

Literature.

Shaler, N. S.—Geology of Cape Ann, Mass. (U. S. Geol. Surv., 9th An. rep., pp. 529-611.)

ARLINGTON MORAINE.

Route.— By electrics (Harvard square car) to Harvard square; change to Arlington car, stopping at Highland avenue, Arlington. Follow this street to the end, where a road leads north to water tower. Strike into the fields on the southwest where, on the hillside, the moraine will be seen.

This is an excessively rocky moraine, composed of medium-sized granite boulders, all of which are well rounded by weathering. The moraine trends generally northeast and southwest, and its northern margin is very strongly defined. It can be traced southwestward for some distance in the Belmont woods. Its appearance suggests a frontal moraine from which the fine material has been washed out, leaving only the coarse blocks.

MYSTIC RIVER QUARRIES, SOMERVILLE.[1]

Route.—*By electrics*, Winter Hill car from Scollay square; stop at Temple street, Somerville; three miles. Quarries on right, near Mystic river, the largest at eastern end of a low hill, which has been largely removed.

By wheel, same as to Medford, stopping on crest of Winter hill at Temple street, and turning to right.

These rocks strictly speaking are not slates, but fine-grained pelites, devoid of good cleavage. They are intersected by numerous sets of joints, large and small, which divide the strata into polygonal blocks of various sizes, often with the regularity of art. It was largely upon specimens from here that Woodworth based his classification of joint-fractures. The beds are cut by at least two series of dikes—an earlier gray set, now deeply altered, and a later dark group of the basaltic type. The former usually extend in a direction nearly east and west, or parallel with the strike of the strata, and send out several sills. One of these, about four feet thick, may be seen in the southwest corner of the easternmost quarry, where it is cut off by the later black dikes. In the northern part of the opening one of the latter shows a remarkable series of included fragments of rocks occurring beneath the Boston Basin. A few annelid or crustacean trails of undeterminable age have been found in the rocks of the quarries facing the Mystic river. Farther west in the open field south of Tufts college this group of pelites is overlain by a small patch of felspathic quartzite, which in turn is capped by a few feet of red shale, the whole being compressed closely into an unsymmetrical syncline, with the steeper side on the east.

Literature.

Woodworth, J. B.—On traces of a fauna in the Cambridge slates. (Bos. Soc. Nat. Hist., Proc., vol. 26, pp. 125-126.)

Woodworth, J. B.—On the fracture system of joints. (Bos. Soc. Nat. Hist., Proc., vol. 27, pp. 163-183.)

[1] The notes for this locality were furnished by Mr. J. B. Woodworth.

COREY HILL, BRIGHTON.

Route.—*By electrics*, Allston and Newton cars from Park street by Subway, alighting at Allston street, Allston. Turn to left up Allston street to Commonwealth avenue extension at foot of Corey hill; to right, up the avenue. The rocks in question extend from the foot of the hill on the north, to the road quarry on the summit.

By wheel, along Commonwealth avenue and its "extension," from Boston to the point mentioned above.

Corey hill is a drumlin, the outline of which has been changed somewhat through artificial terracing. The main hill is sheltered, as it were, on the east of a large projection of sediments over which the extension of Commonwealth avenue runs. The appearance is as though the drift had migrated eastward under the influence of the ice-movement, coming to rest finally in the lea of the great *roche moutonnée*.

The structure of the bed-rock is that of a crushed and faulted anticline, whose axis crosses the road northwestward on the north side of the crest of the hill. The strikes and dips of various outcrops are somewhat obscure. Sandstone, conglomerate, and sandy shale compose the mass, apparently with several repetitions. This appearance may be due, however, to isoclinal folding, and this is known to account for one case. On the east side of the road, the last ledge before the open field on the summit shows sandstone, conglomerate and shale, the whole apparently at least fifty feet thick. Examination of contacts will show that pinched folds are present, giving probably less than six feet total thickness for the mass. In many places cleavage obscures the bedding, lying at a small angle with it. On the west side of the road for some distance it is almost parallel with the stratification, and the two together form the slab-shaped outcrops of the cliff.

Other secondary structures are abundant, and the joint systems are worthy of special notice. The east side of the road, opposite the quarry, gives a face of conglomerate with six or seven systems finely developed, the fracture passing through pebbles and cement with equal ease. On the west of the road and north of the quarry, opposite the large open field, sandstone whose bedding and cleavage coincide shows several good systems, and occasional curious local interruptions to them.

NEWTONVILLE ESKER AND SAND-PLATEAU.

Route.—*By rail*, Boston and Albany road, Kneeland street station, to Newtonville; turn to south, following electric track to Cabot street; fare $.15.

By electrics, Cambridge and Newton cars to Newton, change to Newtonville and Newton Center cars; alight at Cabot street, Newtonville.

By wheel, along Commonwealth avenue and extension from Boston to Newtonville at the electric transfer station; turn to right, to Cabot street. Distance from Boston eight miles. Turn down Cabot street, following it till a gravel ridge crosses it, and the head of the esker lies 200 yards to the left.

The esker can be followed readily southward from its sudden rise from a sand-plain. At one point, however, where it takes a broad sweep eastward, it has been cut entirely away. The relations of cross-section, direction and height of crest-line do not follow very closely the lines recorded by Woodworth for the Auburndale esker. No good cross-section of the esker is visible; unless the gravel pit near Cabot street is in operation. Here at times a very fine anticlinal structure is visible, the core being of cobbles up to five inches in diameter, overlain by an arch of fine assorted sand, with a clean contact between the two. Above this the anticline is unsymmetrical, broadening on the outside of the curve. The esker ends one hundred yards north of the third cross-road met in following it southward, branching into three termini. On the east is swampy land. To the west, the normal topography has been superseded by a secondary sand-plain and two subsidiary eskers starting from near the north end of the first. Kettle holes and kames are abundant in the neighborhood of the esker, especially toward its foot.

South of the terminus of the ridge is a *fosse* extending east and west, beyond which the concave scallops of the head of the sand-plain rise steeply. The material here is coarse, like that of the south end of the esker. At times fresh cuts show the stratification, without, however, rendering visible the "back-set beds" noted by Davis. Walking south to Commonwealth avenue, fine sections can be had near the electric transfer-station, showing fore-set and top-set beds, the latter coarse and separated from the former by an erosion contact. Eolian action often brings out the more resistant laminæ in portions which may be untouched for a few days. To the north, along the electric line, sections of the western lobes are exposed by excavations for a reservoir, and the till beneath the plain laid bare. South of Commonwealth avenue the convex terminal lobes of the plain are developed finely. Still farther south and southwest lies the remnant of a swamp with clay bottom, now largely filled in by human agency. Three hills project above the general level of the plain. They are a drumlin and two kames, older than the plain and partly blanketed by it. The illustration of Gulliver's model shows them well.

Literature.

Davis, W. M.—The subglacial origin of certain eskers. (Bos. Soc. Nat. Hist., Proc., vol. 25, pp. 477–499.)

Davis, W. M.—Structure and origin of glacial sand-plains. (Geol. Soc. Am., Bull., vol. 1, pp. 195–212.)

Gulliver, F. P.—The Newtonville sand-plain. (Journ. of Geol., vol. 1, pp. 803–812.)

AUBURNDALE.

Route.—*By rail,* Boston & Albany road, Kneeland street station, to Auburndale. It is best not to go by wheel on account of the cross-country walking. From the station turn to left along the main street, to right up Grove street (first turn), to left along Woodland avenue, past grounds of Lasell seminary, to right up Seminary avenue, to left up Oak Ridge 100 feet.

The beginning of the esker near the Seminary grounds cannot be seen well. The coarseness of the material in the excavation near the house at Oak Ridge is noteworthy, the presence of two large boulders particularly. The variation in texture from here to the terminus at Woodland is much greater than in the Newtonville case. The cross-section profiles of the ridge at various points are very different, and the changes in this correspond in a general way to changes in height and direction of crest-line as noted by Woodworth, who drew some of his conclusions as to the origin of certain eskers in subglacial tunnels from a study of this ridge. Davis also used the region between Auburndale and Waban in reaching the same results.

The topography of the neighboring surface is normal, the land being swampy, with the exception of some kame and kettle portions in Auburndale. A branch esker a few hundred feet long runs off from the west side of the main ridge between Auburndale and the railroad crossing, evidently the cast of a side tunnel. The cut at the Boston and Albany railroad track gives a cross-section profile unexcelled in symmetry. Farther on, the sudden rise and fall of the crest just before Beacon street is reached are independent of the surrounding topography.

The esker ends about here, which is near the station of Woodland; but its original limits have been effaced by the construction of new roads. Two sand-plateaus are near. On the right, one extends westward to the Charles river, and holds at its southeastern edge the village of Newton Lower Falls. The ice-contact slope has been cut partly away, but ran from north to northwest, on the west side of Beacon street. The frontal lobes extend to the Charles river. The delta is chiefly of interest in having three distinct levels. On the left, and coalescing with the former for a short distance directly along the street, is the Waban plain, which stretches eastward from Beacon street past the Waban station. On its surface, near this station and north of the road between the two points just mentioned, is an interesting kettle hole. The detritus around it grows steadily coarser up to its edge, showing feeding from the melting of a block of ice which was never completely covered. If the railroad track be followed from Waban to Woodland, a large borrow-cut shows one or two features, and once showed many more. At its nearer (east) end, on the south side, the faint outline of an esker can be seen, buried by the Waban plain. Talus has nearly obliterated the view of it. At the farther end of the cut, near Beacon street, is the section of a kettle hole whose ice was buried, as was shown by the inclination of the strata on its sides when the cut was fresh. The ice-contact ran nearly parallel with Beacon street, and the character of the bedding which could be seen in former years showed that no retreat of the ice took place during the formation of the plain.

Literature.

Davis, W. M. — The subglacial origin of certain eskers. (Bos. Soc. Nat. Hist., Proc., vol. 25, pp. 477–499.)

Woodworth, J. B. — Some typical eskers of southern New England. (Bos. Soc. Nat. Hist., Proc., vol. 26, pp. 197–220.)

EXPLANATION OF PLATE

Fig. 1. A. B. map of π-efflux of the Na+ flux variation in the pH of creat-...
changes in the two rates. F. b ... ... ing control;

Fig. 2. Diagrams show the relation between in-seminal A B C D ... ...

Fig. 3. section showing ... between ... ... wampy trans ... ker it bewe D. J. ...
nal ice sheets. R is able T to show

EXPLANATION OF PLATE.

From "Some typical eskers of southern New England," by J. B Woodworth, in Bos. Soc. Nat. Hist., Proc., vol. 26.

FIG. 1. A - B = map of crest-line of the Auburndale esker. C - D = variation in height of crest-line, showing relations between changes in the two lines. E - F = general level of the surrounding country

FIG. 2. Diagrams showing relation between probable cross-section of ice-tunnel (A B C D) and present cross-section of esker (E F G).

FIG. 3. Section showing relation between esker and esker-tunnel (C), swampy areas or kettle holes (D), hillside kames (E), and residual ice blocks (B) beside the esker.

1

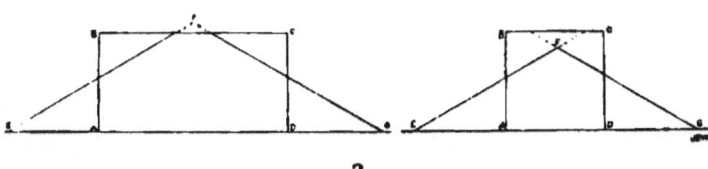

2

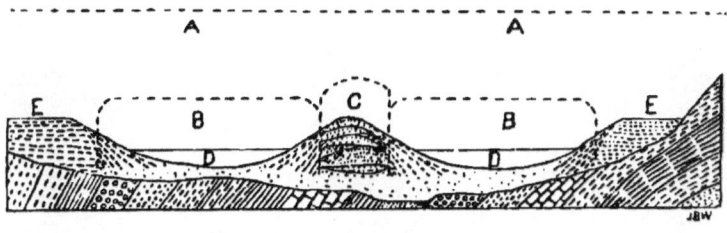

3

GEOLOGY: SOUTH SHORE.

Prof. W. O. Crosby.

THE OUTER ISLANDS OF BOSTON HARBOR.

Lovell's, Gallop's, George's and Great Brewster islands are drift-covered. The rest have little or no glacial detritus, and form part of a great synclinal fold of slate, with intrusions of diabase, in part sills. Calf island lies on the north side of the fold, and the Brewsters on the south. The south side of Middle Brewster island gives the best exposures of alternating diabase and slate. The former possesses a very perfect flow-structure, and shows concentric weathering on a large scale.

NANTASKET AND COHASSET.

Contemporaneous flows, dikes, faults, plutonic rocks, etc.

Route.—By Nantasket steamers from Rowe's wharf, Boston, to Nantasket direct, fare, $.25; or to Pemberton by same steamers and thence by the Nantasket Beach railroad to Nantasket station, or by all rail from the Old Colony station (Kneeland street) to Nantasket station, fare $.35.

The district in the southern part of Hull known as Nantasket, and the adjacent shore of Cohasset, embrace in a limited area many interesting and instructive outcrops, and afford an abundance of material for two excursions. The rocks, above the granitic series, are chiefly conglomerate of probable Carboniferous age interstratified with many contemporaneous flows of basic and neutral lavas (melaphyr and porphyrite) and beds of tuff; and the whole is intersected by several systems of dikes.

1. Nantasket beach to Green hill.

This coastal area of Nantasket can be studied to the best advantage when the tide is out.

Commencing at the southern end of Nantasket beach we find in Long Beach rock, which projects into the sea from the base of Atlantic hill, a bed of conglomerate overlain by (1) a thin layer of finely laminated greenish tuff of jaspery hardness; and (2) a very complete and typical flow of melaphyr some sixty feet in thickness. This flow is dense and crystalline in the lower and central portions, while the upper part is highly scoriaceous and shows fluidal lines; the whole recording a quiet submarine or littoral eruption which was preceded by explosive action projecting into the water a limited amount of impalpably fine dust. Ascending Atlantic hill from the beach we find that this eruption was followed by a series of eruptions which were alternately quiet and explosive, forming flows of melaphyr and beds of tuff and agglomerate; the whole being capped by a flow (possibly composite) three or four hundred feet in thickness, which became very generally brecciated by continuing to flow after it began to harden.

The west side of the small beach between Atlantic and Central hills presents a good section of this volcanic series. One of the flows exposed in this section has preserved very perfectly its original wavy or undulating surface; while another, which expands rapidly eastward, encloses many bomb-like masses of amygdaloidal melaphyr. The great bed of melaphyr of Atlantic and Central hills is continued southward in Willow Ledge hill, where it encloses a very typical bed of tuff; and eastward in Gun Rock, several half-tide ledges, and the northernmost of the Black Rock islets. The principal Black Rock islet and several neighboring ledges consist of porphyrite of decidedly felsitic character, and probably mark approximately one of the ancient volcanic vents of the Nantasket region.

The diabase dikes of this shore present some points of special interest. They belong chiefly to two approximately east-west systems, those trending north of east being the older, as proved by a very clear intersection on the shore east of Gun Rock; and dikes of both these systems are cut by a north-south dike in the conglomerate at the northern base of Green hill. The largest of the three dikes on Long Beach Rock outcrops again on the shore near Gun Rock as a composite dike, consisting of six parallel branches. Several dikes are well exposed on Gun Rock. One of these cuts without faulting a remarkably regular quartz veinlet; and encloses in porphyritic fashion minute fragments and single crystals derived

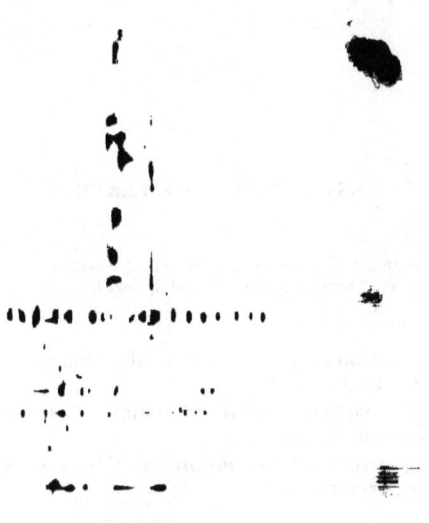

EXPLANATION OF PLATE.

From "Geology of the Boston Basin" by W. O. Crosby, in Bos. Soc. Nat. Hist., Occasional Papers, vol. 4, part 1.

FIG. 1. Natural section of the inclined double dike (34) on the western end of Rocky Neck.

FIG. 2. Dike 157 cutting a ledge of granite, near White Head, on Hominy point, Cohasset.

FIG. 3. A highly brecciated portion of the thin bed of indurated tuff on Long Beach rock.

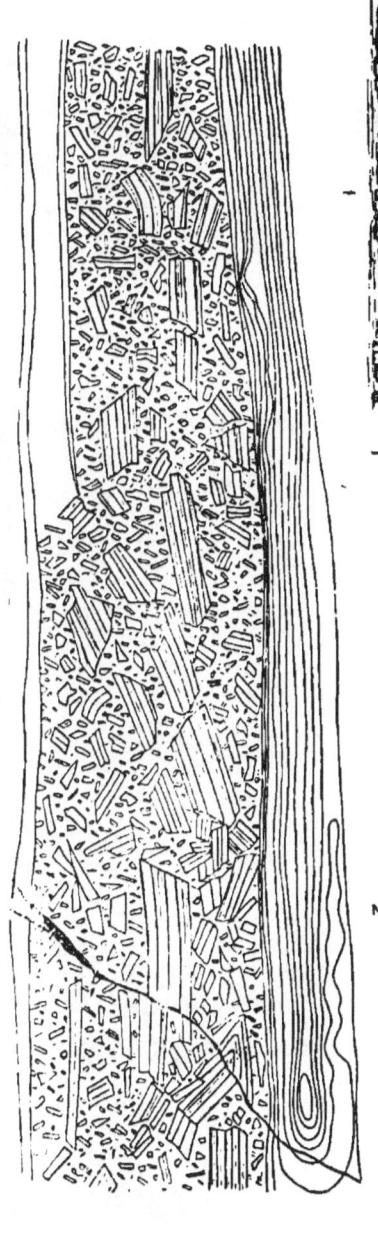

1

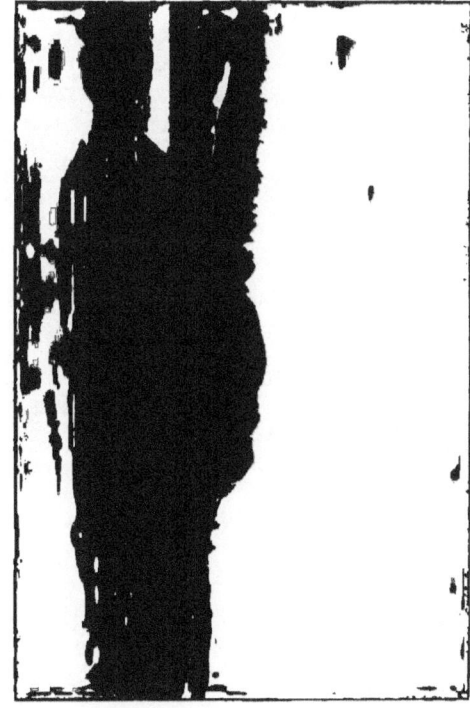

2

from the granite, which is thus proved to underlie the volcanic rocks and conglomerate.

2. *The Nantasket ledges west of Hull street.*

Route.—The chief points of interest in this area are reached most readily from Nantasket station by walking south along the railroad to Great hill on the extreme south side of the harbor.

As we approach Great hill, which is the bold and isolated eminence on the west of the track, we have on the east side the beginning of an interesting north-south section, extending over Crescent hill (opposite Great hill) to Marsh island in the Weir river marshes. This section consists, from below upward, of the Third Conglomerate, Second Melaphyr, Fourth Conglomerate, and Third Melaphyr. The contacts are admirably exposed and unmistakably contemporaneous. The section is broken by numerous east-west faults, repeating the beds to a considerable extent; and this faulted zone is bounded on both east and west by important north-south faults. The easternmost fault separates it from Melaphyr plateau, in which these two beds of basic and amygdaloidal melaphyr are situated, brought together in one broad and continuous outcrop.

We next cross the marsh westward from Great hill to Granite plateau and the shore of Weir river. Scattered over the granite, which is clearly the fundamental rock for this stratified series, are patches (outliers) of the first or basal conglomerate; and the roots of this conglomerate may be seen filling original cracks and fissures in the granite, and forming one type of sandstone dikes. Proceeding north from Granite plateau to Granite point and Cliff plateau, the basal conglomerate is found to be overlain by the First Melaphyr, a compact, non-amygdaloidal variety, quite distinct in character from any other Nantasket flow. It is characterized in part by segregations of bright red jasper; and this feature is common also to the underlying and overlying First and Second Conglomerates. The section is particularly clear in West Porphyrite hill, northwest of Cliff plateau, where we have in regular sequence, with the contacts well exposed, the First Melaphyr, Second Conglomerate and a heavy bed of porphyrite. North of this hill, and separated from it by a profound fault, is Melaphyr peninsula, affording an exceptionally fine section of a typical but composite lava flow. This melaphyr should probably be correlated with the Second or Third Melaphyr east of the railroad.

Northeast of Cliff plateau, the section of West Porphyrite hill is exactly repeated in East Porphyrite hill, except that we find above the porphyrite bed traces of the Third Conglomerate, which has a more complete development immediately to the south, in Conglomerate hill. Great hill shows granite on the south overlain by the basal conglomerate and First Melaphyr; but these are cut off abruptly by a fault which lets down the Third Conglomerate to form the main body of the hill.

Southwest of Great hill, on the border of the marsh, the granite is cut by a large dike of melaphyr identical with the First Melaphyr and believed to be the vent through which that flow reached the surface. The diabase dikes are similar to those of the coastal area, two prominent systems of east-west dikes being cut by a north-south system; and the finest example of the latter is a composite dike on the shore between Cliff plateau and East Porphyrite hill.

Rocky neck, in Hingham, on the west side of Weir river, is geologically very similar to the western area of Nantasket, but it is less accessible.

3. *The Cohasset shore.*

This excursion embraces the shore ledges along Jerusalem road from near Green hill, Nantasket, to Cohasset.

Route.— The best route is by boat or rail to Nantasket station, and thence by electric cars to Jerusalem road near the Black Rock house. The phenomena of most particular interest may be observed between Green Hill beach and Pleasant beach (one and a half miles), whence the return can be made on foot or by barge back to the Black Rock house and Nantasket; or the walk may be continued to a point just beyond the bridge across the mouth of Little harbor and thence by the right hand road to Cohasset village (four miles in all), returning by train to Boston.

No locality near Boston offers a finer field for the study of plutonic rocks and dikes. The former embrace: (1) the normal biotite granite (granitite); (2) the basic or dioritic granite, passing into (3) diorite; and (4) fine-grained and highly acid granite, passing into micro-granite and quartz porphyry.

Numbers 1 and 2, the latter especially, are in part distinctly gneissoid (flow structure). Number 1 cuts 2, and both cut and enclose the diorite, in which the gneissoid structure also often is strongly marked. The fine granite (4) forms narrow branching dikes, cutting all the other types. All are regarded as successive phases developed during the cooling and crystallization of one magma; and the diorite is believed to occur largely but not exclusively as segregations in the basic granite.

The dikes embrace, in order of age: (1) porphyrite dikes, which have a general north-south trend and are believed to radiate from the great mass of porphyrite forming the Black Rock islet and probably marking the vent for this type of lava; (2) the three systems of diabase dikes noted at Nantasket. Some very clear and interesting intersections are exposed. If this excursion is extended to Cohasset village, an hour should be reserved for a visit to the glacial pot holes on Cooper's island (really a peninsula) in Little harbor. The beautiful sloping sand-plain on which the village stands, with its bold ice-margin toward Little harbor, and fine kettle holes, one of which is called the Punch-bowl, also should be noted.

Literature.

Crosby, W. O.— Geology of the Boston Basin. (Bos. Soc. Nat. Hist., Occasional Papers, vol. IV, part 1, 1893.)

NORTHERN HINGHAM.

Route.— This district may be reached by the Nantasket steamers to Downer Landing and Hingham, or by the South Shore trains of the Old Colony railroad from the Kneeland street station.

The areas of sedimentary rocks in the northern part of Hingham present several good sections of the Carboniferous strata of the Boston Basin, contrasting with and supplementing the Nantasket sections. The volcanic rocks are less abundant; but there is above the granitic rocks one heavy bed of melaphyr followed by many alternations of conglomerate and red slate, and this conglomerate series is conformably overlain by a great thickness of gray slate, while plication takes the place of faulting as the dominant type of disturbance or secondary structure.

The most accessible section is that at Downer Landing, in Melville Garden and adjoining fields, and on the islands of Hingham harbor (rowboat from Melville Garden), a gracefully curving monocline of alternating conglomerates and finer sediments, with several very obvious transverse faults.

A mile to the westward across the fields is Huit's cove, where a great development of melaphyr is bordered by conglomerate, important strike faults intervening; and the conglomerate is overlain conformably by more than a thousand feet of slate. A part of the conglomerate contains pebbles and boulders of limestone, probably derived from the Cambrian series of this region.

From Squirrel hill southwestward across Beal street to Beal's cove we have a practically complete section from the granite and felsite up through the entire conglomerate series and five hundred feet into the overlying slate series; and in Hingham village, on and near the line of Hersey street, this section is repeated, but the beds are inverted and the great slate series cut out by a strike fault.

Aside from the melaphyr, the only igneous rock of this area meriting special attention is the red felsite, a truly effusive type, characterized by fluidal and brecciated structure, of which many handsome boulders may be seen in the stone walls on Lincoln and Thaxter streets, on the north side of Bradley hill. Ledges of the red felsite formerly existed on the line of Lincoln street, but were obliterated in the grading of the road. The unique character and limited area of this rock have made it useful in tracing the distribution of the drift of the region.

The area traversed in this excursion embraces a number of very typical drumlins and some good developments of the sand-plains marking the lower levels of glacial lake Bouvé.

Literature.

Crosby, W. O.—Geology of the Boston Basin. (Bos. Soc. Nat. Hist., Occasional Papers, vol. IV, part 2, 1894.)

MILL COVE, NORTH WEYMOUTH.

Outcrop of lower Cambrian strata.

Route.—Train from Old Colony station (Kneeland street) to Quincy electric car from Quincy to North Weymouth, thence walk south on Pearl street, about half a mile; or by train to Weymouth Highlands, and thence walk northwest about half a mile.

The red Cambrian slates outcrop in the road; but the best exposures are one-fourth mile to the west, on the wooded point which is partly isolated by the salt marsh at the north side of the cove. About seven hundred feet of strata are exposed, mostly red slate, with several thin beds of white crystalline limestone and innumerable calcareous streaks and nodules. Characteristic lower Cambrian fossils have been found, but are not plentiful.

On the south side of the cove is granite, and in this direction the calcareous portions of the slate show interesting alteration. The chief secondary mineral is epidote, occurring commonly as a shell enclosing the more purely calcareous center of the nodule which, on the weathered surface, usually has been removed by solution, leaving the epidotic shell hollow.

To the north of Mill cove is a finely moulded frontal slope of one of the deltas (Quincy plain) of lake Bouvé. Between Weymouth Highlands and Weymouth station (one and one-half miles) the railroad passes through two long cuts in the dark middle Cambrian slates, with exceptionally good exposures of the fault contact between the slate and granite.

THE MONATIQUOT VALLEY, HAYWARD CREEK (PARADOXIDES QUARRY) AND RUGGLES CREEK.

Cambrian strata and their relations to the granite, including the only known localities in the Boston Basin for middle Cambrian fossils.

Route.—Train to Weymouth station, electric car from Weymouth *via* Hayward creek to Quincy, and thence by train or electrics to Boston.

Monatiquot river becomes Weymouth Fore river at the head of tide water. This valley has a floor of middle Cambrian slate and walls of granite, with a fault contact on the south side and an original igneous contact on the north side. The fault contact is exposed well in the railroad cut immediately east of Weymouth station. West of the station, on Quincy avenue, the granite and slate are separated by a diabase dike more than two hundred feet wide. Crossing the valley obliquely through Shaw street and returning eastward to Quincy avenue on Allen street, north of the river, the igneous contact of the slate and fine granite may be observed at several points ;. and at one point about half way from Shaw street to Quincy avenue several dikelets of fine granite penetrate the slate. *Paradoxides harlani* has been found by Mr. T. A. Watson on the north bank of the river about half a mile east of Quincy avenue, and south of the river on Commercial street near Liberty street.

Taking the electric cars on Quincy avenue, we cross Wyman

hill, a ridge of fine granite a mile wide, to Hayward creek. The Paradoxides quarry is on the south bank of the creek at its mouth and one-third of a mile east of the avenue. A few rods southeast of the quarry the igneous contact of the slate and fine granite is exposed obscurely in the marsh, as described by Wadsworth. This contact runs in a west-northwest direction to the creek, and north of the creek it may be traced in the same direct line, with numerous exposures over Eldridge hill. On the north side of Eldridge hill the slate ends abruptly against a large trap dike which undoubtedly occupies a fault fissure. These phenomena are substantially repeated, without the dike, in the valley of Ruggles creek, north of Eldridge hill; a long narrow belt of slate meeting the fine granite on the south in an igneous contact (actually exposed only at a few points), and on the north in a remarkably straight, continuous and conspicuous fault scarp a mile long.

From Ruggles creek, Quincy may be reached by electric cars on Quincy avenue or on Washington street half a mile farther north.

BLUE HILLS OF MILTON.

Relations of the Quincy granite to the quartz porphyry and other forms of aporhyolite (felsite).·

Route. — Train to Quincy Adams station, by New Haven road, Kneeland street station, returning from this station or West Quincy; fare, $.17.

This excursion involves a rough walk of five or six miles, but is replete with interest for the student of granitic rocks. It can be made advantageously only with a guide, on account of the wilderness-like character of the area; and hence need not be described in detail here.

QUINCY GRANITE QUARRIES.

Route.—Train to West Quincy, by New Haven road, Kneeland street station; fare, $.15.

The principal quarries are on the hill immediately west of the railroad, extending back from one to two miles, and on North Common hill nearly a mile east of the railroad. Among the features of special geological interest are the joint structure — both block and sheet quarries being well represented — the relations of the gray and black granites, and the intricate relations of the granite to the Cambrian slate on the north slope of North Common hill.

LAKE BOUVÉ.[1]

(An extinct glacial lake.)

Route. — Boston to Quincy by New Haven road, Kneeland street station; fare, $.15. From Quincy by electrics or wheel.

Taking the East Weymouth car at Quincy and leaving the village, one can have a fair view of the "Quincy plain" of lake Bouvé, which has an elevation of forty feet. It is cut in two by the Weymouth Fore river. The car continues on this plain to Old Spain, where it turns south and soon descends to the marsh lands, passing down the delta front of the plain. As the car descends look west to see the front of the plain along the Fore river. Less than a mile south of this plain, the road passes through a cut in the narrow esker-like portion of a small representative of the "Hingham plain" (fifty feet). To the west of the road it is narrow and looks like an east and west esker. A good section is exposed, where it is cut by another road. This section when not covered by talus shows stratification and ripple-marks well. To the west of the main road the plain is broader. A large kettle hole occurs in it. Its northern end is an ice-contact slope, of high grade in most places. Its southern margin shows a steep but well-marked lobate "delta front." Follow this plain west, to Weymouth Fore river, where a series of north-south eskers occurs. From the top of King Oak hill, a drumlin just south of Weymouth Heights station, a comprehensive view of this and some other parts of the extinct lake can be obtained. Taking the car again for East Weymouth, and changing to the South Weymouth car, one passes for a long time over a part of the "Whitman plain" (one hundred feet). This encloses Whitman pond. The plain can be seen well only by leaving the car at Lovell's corners and penetrating the woods. Just east of the pond occurs a large rocky tract which rises island-like above the plateau. Several outliers of this plain occur to the east and west of the pond.

Passing southward from Lovell's corners, leave the car at the fork of the road, and take the eastern branch, ascending the north-

---

[1] The material for this section was furnished by Mr. A. W. Grabau.

ern slope of "Liberty plain" (one hundred and forty feet) where it is used as a cemetery. Follow the western margin and note that it is a steep lobate delta front, not an ice-contact slope. Southward can be seen the valley of Old Swamp river, one of the outlets of the lake at the Liberty stage. The road leads east across Liberty plain, and shows many rocky shore lines. Leaving the plain go to Queen Ann's corners, where the car may be taken for Hingham. The car passes northward over the eastern lobe of Liberty plain, across a *fosse*, and on to Glad Tidings plain (seventy feet). Leave the car at the first cross-road after reaching this plain, and take the road leading east, continuing for two miles. Passing through a belt of moraine, the road crosses one of the highest hills of this moraine, situated about a mile north of Prospect hill. At the highest point, climb the fence and descend the northeastern slope, reaching at eighty feet below its summit a well-marked cliff cut into the northern margin of the hill. North of this cliff the valley floor is level, and thickly strewn with boulders. This is believed to be the outlet of the Whitman stage of lake Bouvé, whence the waters passed southeastward.

To return, take the car for Hingham, passing across the eastern lobe of "Glad Tidings" plain, and finally onto and over "Hingham plain" (fifty feet). At Hingham the train for Boston can be taken.

The western part of the lake, including Monatiquot bay, cannot be visited readily without a map or guide, and requires more time.

MARTHA'S VINEYARD AND GAY HEAD.

Route. — To Cottage city from Boston by rail and boat connections, Old Colony system, Kneeland street station. By stage to Gay Head.

All that can be done here is to refer to the literature on the subject, on account of the size of the area involved.

Literature.

Shaler, N. S.—Geology of Martha's Vineyard. (U. S. Geol. Surv., 7th Ann. rep., pp. 297-363.)

Shaler, N. S.—On the occurrence of fossils of Cretaceous age on the Island of Martha's Vineyard. (Harv. Coll., Mus. Comp. Zoöl., Bull., vol. 16, pp. 89-97.)

Shaler, N. S.—Tertiary and Cretaceous deposits of eastern Massachusetts. (Geol. Soc. Am., Bull., vol. 1, pp. 443–452.)

Shaler, N. S.—Pleistocene distortions of the Atlantic seacoast. (Geol. Soc. Am., Bull., vol. 5, pp. 199-202.)

Woodworth, J. B.—Note of the occurrence of erratic Cambrian fossils in the Neocene gravels of the island of Martha's Vineyard. (Am. Geol., vol. 9, pp. 243-247.)

Woodworth, J. B.—Unconformities of Martha's Vineyard and Block Island. (Geol. Soc. Am., Bull., vol. 6, pp. 197-212.)

GEOLOGY: TURNER'S FALLS REGION.

TWO EXCURSIONS IN THE CONNECTICUT VALLEY.

Prof. B. K. Emerson.

1.

Start in the morning from the Mansion house in Greenfield, and drive five miles across the sand-plain to Leyden Green, where is a very fine and instructive contact of Triassic sandstone upon the Leyden argillites showing the far-travelled character of the arkose. Drive thence four miles across the strike of the sandstone to the exposure of the basal contact of the trap sheet on the sandstone, showing the frothing up of the sand into the trap, and many curious contact phenomena. Among these is to be noticed especially the formation of diabase pitchstone, as described by the writer. Climb to the lookout over the valley, then drive east to the upper contact of the sandstone on the trap, which is coarsely amygdaloidal and full of minerals.

A visit to Turner's Falls, the fault at the mouth of Fall river, and the chlorophæite locality, is full of interest. Drive two miles east to the "bird-track" quarries at the Lily pond, where is an abandoned waterfall of the Connecticut with its cañon in which lies Lily pond. Return to Greenfield by electrics from Turner's Falls. (See also Palæontology, p. 43.)

2.

Go from Northampton in the morning by train (five minutes) to Mount Tom station. Cross in the ferry and see Titan's piazza, overhanging trap columns with sandstone contact; and Titan's pier, with the contact of underrolled trap full of limestone inclusions and amygdaloidal at the base. Recross and stop the train at the north line of Holyoke, to see the trap at Larrabee's quarry, where mud has been kneaded into the trap at the surface; and to look at the tuff beds. Then cross the whole Triassic series to the top of Mount Tom, seeing remarkable intruded dikes and sills; the surface of the trap sheet where ejected fragments have fallen on the lava sheet while it was moving; an excellent exposure of the great fault which cuts through the Mount Tom range; and the largest *Brontozoum* tracks *in situ*. The ascent of the mountain will be made by electrics, and the view is in many respects the finest in the valley. From the top of the mountain one can ride by electrics to Holyoke or Springfield.

Literature.

See papers especially by B. K. Emerson; the one referred to above being *Diabase pitchstone and mud enclosures of the Triassic trap of New England.* (Geol. Soc. Am., Bull., vol. 8, pp. 59-86.)

SURFACE DEPOSITS.

In the Greenfield excursion described above, the drive will take one across the filled-up northern lobe of Hadley lake; and at the northwest corner opportunity will be given to see where an arm of the glacial sheet projected into the lake, preventing the filling of a portion near the shore. One can study also a fine dissected delta thrust into this depression after the ice had melted away.

From the lookout on the trap ridge east of Greenfield, one can follow the lake beach southward and see the excavation made in it by the Green river, and the broad well-terraced depression cut by the Deerfield river. To the east across the Connecticut will be seen the front of the broad sand-plain of the delta of Miller's river, which extends six miles eastward. In Turner's Falls the river has worn fine terraces in these beds, exposing the clays, and has cut back a postglacial rock gorge and made a good waterfall. At Lily pond is also an excellent specimen of an abandoned waterfall and cañon, where the Connecticut has worn round a sandstone ridge and forsaken its old course.

At Northampton drumlins are developed well on both sides of the valley. The shore deposits of Hadley lake are shown well and the Champlain clays exposed at the asylum, and exhibit complex contortions from the stranding of glacial icebergs. The ox-bows, meadows, and new islands of the Connecticut are seen to advantage, and the terraces are exceptionally fine south and west of Northampton, and at Florence.

An excursion may be taken by carriage five miles east of Amherst for the study of perfectly preserved high-level glacial lake beds and a grand serpent kame. The Pelham "asbestos mine"— an olivine-enstatite core with a contact margin containing asbestos, anorthite, biotite-corundum and many other minerals — can be seen.

PALÆONTOLOGY: EASTERN MASSACHUSETTS.

Amadeus W. Grabau.

Fossiliferous rocks, belonging to the following geologic formations, are found in eastern Massachusetts.

    Recent.
    Neocene.
        Pliocene.
        Miocene.
    Eocene (Erratics).
    Cretacic.
        Upper? Cretacean.
        Middle? Cretacean.
    Jurassic?
    Triassic.
    Carbonic.
        Middle Carbonian.
    Devonic.
        Upper Devonian.
    Cambric.
        Upper Cambrian (Erratics).
        Middle Cambrian.
        Lower Cambrian.

It will be seen that, of the larger rock series, only the Permic, the Siluric and Ordovicic are unrepresented by fossiliferous beds, although probably the last two occur in this vicinity. The Jurassic of Martha's Vineyard is regarded by palæobotanists as lower(?) Cretacean on account of its plant remains. The acceptance of the Jurassic age of these beds rests at present upon their identification by Professor Marsh with others of the Atlantic coast in which he has found Jurassic Dinosaurs.

The Cambric and Recent beds are represented best in the immediate vicinity of Boston, but occur in many places in eastern Massachusetts. The Neocene, Cretacic and Jurassic (?) are confined to the South Shore islands, while the Eocene is found only as transported material in the drift plains of Truro, Cape Cod. The Triassic is confined to the Connecticut valley. The Carbonic is found in many places in eastern Massachusetts, but its fossils are few. The Devonic has so far been identified definitely in only one locality on the Connecticut river,—*i. e.* Bernardston, Franklin county.

CAMBRIC SERIES.

A. Lower Cambrian beds.

1. EAST POINT, NAHANT.

Route.— From Lynn (see pp. 10–11) by barge or bicycle across Nahant Neck to the Lodge estate; follow footpath to cliff on the left; barge fare, $.25 round trip.

The stratified rocks of East point, Nahant, consist of altered slates and limestones of Eocambrian age, as first definitely ascertained by Foerste. The limestone beds furnish fossils which, up to the present time, number only a few species in all. They are chiefly species of Hyolithes, and appear usually as cross or longitudinal sections on the rock surfaces. In the former case they exhibit more or less distorted rings, according to the direction of the section; while in the latter case they are marked by lines, either parallel or converging. The shells differ slightly in color and texture from the enclosing rock, and on this account can be seen on the sections. As a rule, however, it takes some time before one becomes accustomed to recognizing these minute markings. Complete specimens can be obtained only by breaking the rock carefully with a hammer; and considerable time and patience are necessary to insure a fair reward.

A common species is *Hyolithes communis,* var. *emmonsi,* characterized by the semicircular outline of its cross-section, the flattened side occasionally becoming concave. *H. communis,* a more slender form with a nearly circular cross-section, occurs frequently. *H. princeps,* a large species, often with sub-semicircular cross-sections half an inch or more in diameter, is not infrequent; and *H. impar* with a regularly oval section also occurs. In addition to these, imperfect specimens of Iphidea, Stenotheca, and other forms have been found.

2. MILL COVE, NORTH WEYMOUTH.

Route. –By train, Old Colony branch, N. Y., N. H. & H. railroad, Kneeland street station, to Weymouth Heights, fourteen miles; single fare, $.25. Or, by train (same station) to Quincy, thence by electrics or bicycle to Weymouth Heights. Walk north and northwestward around border of Mill cove. See p. 26.

The rocks exposed here are red sandy slates and calcareous strata with some white limestone beds. Fossils are scarce. Some obscure traces of Hyolithes have been found, however; and trilobite remains, probably of the genus Agraulus, have been noticed. (Crosby.)

3. NORTH ATTLEBOROUGH.

Route.—By train, from N. Y., N. H. & H. railroad, Park square station, to North Attleborough; fare $.75. Take electrics at North Attleborough square for South Attleborough, stopping about half-way where, from the apex of a triangle made by the roads, a road leads west to Hoppin hill. In the center of the triangle is a house, which will serve as a landmark. Follow Hoppin hill road to foot of hill, where it crosses a small brook. Shaler and Foerste's locality 1 is just to the right of the road on the banks of the brook. Locality 2 is farther up the brook.

The Cambrian slates exposed here are much disturbed, and involved in a rather complicated series of folds. Nevertheless the fossil-bearing strata are not affected very strongly, and the remains are mostly in a good condition. Hyolithes is the predominating fossil, five species having been described by Shaler and Foerste. Hyolithellus and Salterella also occur. The rocks in places are made up of the shells of the Hyolithes; and wherever the surface is weathered, rather perfect specimens may be obtained. In the more shaly portion of the rock, the shells have frequently a dark color which causes them to appear prominently on the broken surfaces. Trilobites are represented in these slates by the genera Microdiscus, Olenellus, Agraulus and Ptychoparia. The gastropoda include a Scenella, several species of Stenotheca, a Platyceras and a Raphistoma. The occurrence of pelecypods is doubtful, but brachiopods are represented by two species of Obolella, *O. crana* and *O. atlantica*. (For descriptions and illustrations of this fauna see paper by Shaler and Foerste noted below.)

B. Middle Cambrian beds.

BRAINTREE QUARRY.

Route.—By train, from Old Colony station, Kneeland street, to Quincy, seven and three-fourths miles; single fare, $.15. Thence by electrics or bicycle to Quincy point, turning south just before reaching the bridge across the Weymouth Fore river. Follow the road, crossing Ruggles creek, for one mile to Hayward's creek. Follow right bank of Hayward's creek to quarry.

This is the famous Braintree trilobite quarry, from which the first authentic Massachusetts trilobite was obtained, in 1856; although the species had been described as early as 1834 from a specimen of unknown source.[1] The trilobite is *Paradoxides harlani* Green; and since the finding of the first specimen, numerous more or less perfect individuals have been obtained, the best of which are now in the collection of the Boston Society of Natural History. Fragments of this species are common on the loose blocks in the quarry, but good specimens can be obtained only by blasting. Two other trilobites have been found here, *Agraulus quadrangularis* Whitfield, with a smooth, more or less squarely truncated glabella, and *Ptychoparia rogersi* Walcott, with the glabella rounded anteriorly and the frontal limb with a well marked marginal rim. Two pteropods are known, *Hyolithes shaleri* Walcott, and *Hyolithes ? haywardensis* Grabau. Both are rare. The former is recognized by its large size, rapidly tapering form, and biconvex transverse section; while the latter is characterized by a slender, slightly curving form, and nearly circular cross-section. *Parmophorella acadica* Matthew (=*Discina acadica* Hartt) also has been found in these argillites, but so far only poor specimens have been obtained.

C. Upper Cambrian beds.

Quartzite pebbles containing Scolithes and Lingula not infrequently are found on the south shore and on Martha's Vineyard. They are derived commonly from the Carbonic conglomerates.

DEVONIC SERIES.

These rocks, though probably well represented in eastern Massachusetts, nevertheless have been found to contain fossils in only one locality, in Franklin county.

[1] This specimen is now in the collection of the Boston Society of Natural History.

BERNARDSTON.

Bernardston lies on the Boston and Maine railroad a few miles west of the Connecticut river, and is the last village this side of the Vermont line. The locality of the crinoidal limestone is about three-fourths of a mile north of the "New England house" on the east slope of West mountain, back of the house of Mr. Williams. (See Dana's map, '73.) The exposures are in a number of "pits."

The fossiliferous rocks consist of limestones overlain by quartzites, the latter having a shaly character just above the limestone. The fossils of the quartzite occur chiefly in these shaly layers.

Whitfield regards the limestone as of middle Silurian age; and the shaly beds overlying, as of middle Devonian age.

In the shaly layers above the limestone, occur, according to Whitfield, many casts and impressions of brachiopods, including:

Strophomena rhomboidalis (= *Leptœna rhomboidalis*).
Spirifer cf. *disjunctus*.
Rhynchonella, two species.
? *Cryptonella eudora*.
Cyrtina cf. *hamiltonensis*.

A number of other species occur, and also some corals (Streptelasma?).

In the limestone, Whitfield found two species of Favosites, Crinoid stems, and a Syringopora?.

In his recent paper, Prof. B. K. Emerson concludes that the "limestone, magnetite, and the base of the quartzite above the limestone, may be placed with certainty near the base of the Chemung" (p. 374). This makes the whole series upper Devonian.

CARBONIC SERIES.

Rocks of the Carbonic series occur in many places in eastern Massachusetts; but few localities are fossiliferous, and those that are seldom furnish fossils in a well-preserved condition. Probably the best-preserved fossils found in this region were those obtained by Teschemacher from the old anthracite coal mines at Mansfield (N. Y., N. H. & H. railroad, Providence div., 25 miles). He mentions the following species:

Sphenopteris dubuissonis Brongniart.
Pecopteris cisti Brongniart.
P. serlii Brongniart (=*Alethopteris serlii*).
P. punctulata.
Neuropteris cf. *acutifolia* Brongniart.
N. cf. *sechuchzeri.*
Odontopteris sp.

More recently, well preserved ferns have been found in the ledges of Carbonian slates between Mansfield and Attleborough.

Joseph H. Perry in 1885 reported the finding of two specimens of Lepidodendron in the mica-schist which surrounds the granite knoll in the eastern part of Worcester, and which furnishes the Worcester coal. These were identified by Lesquereux as the rare *Lepidodendron (Sagenaria) acuminatum* Goeppert.

C. T. Jackson in 1851 reported the finding of Calamites at Bridgewater, Mass.

The following two localities are the only ones readily accessible for the study of our Carbonic fossils.

ROCKDALE (PONDVILLE STATION).

Route.—By rail, N. Y., N. H. and H. railroad, Providence div. (Park square station) to Pondville station, 23.5 miles; fare $.55.'

This is Crosby and Barton's original locality for Carbonic fossils in the Norfolk County basin. The fossils occur in a small-pebbled or arenaceous conglomerate lying near the top of the first or conglomerate series. They consist of moulds of Sigillaria and Calamites. The moulds are all much compressed, the flattening corresponding in direction with the imperfect cleavage of the rock. The smallest found by the authors was one or two inches in diameter, the largest twenty by six inches on the cross-section. These hollow moulds are usually inclined to the horizon, and some of them have been probed to a depth of twenty feet or more. About thirty specimens were reported by the authors.

J. B. Woodworth in 1894 examined and confirmed the organic origin of these moulds. He reports that, between the casts (whenever such is present) and the mould in the sandstone matrix, there occurs a "limonitic and cellular layer, apparently representing the cortical part of the plant, too poorly preserved for identification." This layer was observed in many of the hollow moulds.

CANTON JUNCTION.

Route.— By rail, N. Y., N. H. & H. railroad (Park square station) to Canton Junction, 14.5 miles; fare $.30. Walk back on railroad track, to rock cut, about half a mile north of station.

The fossils found here were mentioned originally by W. W. Dodge, in his "Notes on the geology of eastern Massachusetts."[1] He speaks of them as "dark cylindrical forms which appear to be branching stems of some kind." According to Woodworth, the fossiliferous beds belong to the "Carboniferous gray series," and appear about 2500 feet north of the granitite exposure of the cut. The rocks are "greenish and grayish slates, originally fine muds, with interlaminated sandy layers. The strike is about N. 60° E. and the dip 65° to 70° S.; the cleavage dips north about 70°" (Woodworth). The plant remains occur in the fine shaly partings between the sandy layers. They are Calamites (cf. *C. aistii* Brongniart) and Sigillaria. In some of the lower slaty and sandy beds, exposed farther north, occur stems of ferns and Calamites impressions.

The most recent discovery of Carbonic fossils in this region is by Mr. M. L. Fuller, who found plant remains considered to be Calamites and probably Sigillaria in arenaceous and conglomerate beds, in the railroad cut about a quarter of a mile northeast of the railroad station at Brockton (N. Y., N. H. & H. railroad, Plymouth div., Kneeland street station, 20 miles, fare $.45.) Fuller states that "in most of these cases the bark has been changed to a layer of highly ferruginous anthracite, surrounding a core of sand" (p. 197). In a few cases the entire trunk has been preserved ". . . as a black, fibrous substance, closely resembling ordinary charcoal." This is friable, and consists of a mixture of carbonaceous material with spicules and fibres of a whitish substance, which on analysis proved to be calcium carbonate. This constitutes over eighty percent of the mass.

TRIASSIC.

To study the Triassic formation and its fossils, the Connecticut valley must be visited. (See Emerson's excursions in the Connecticut valley, p. 33.) The most prominent organic

---

[1] Proc. Bost. Soc. Nat. Hist., vol. 17, 1875, p. 414.

remains in these rocks are the footprints, fishes and plants. The footprints are found in many places. The first described by Hitchcock came from Montague, Gill, Northampton (east side of Mount Tom), and from South Hadley. Numerous localities have been discovered since, Hitchcock giving thirty-eight in his great memoir. The most prolific as well as most important is probably Turner's Falls, and the vicinity of Lily pond.

Fossil fishes occur plentifully at Turner's Falls, and they have been found also at Chicopee Falls, Amherst, Hadley Falls, Sunderland, Deerfield and West Springfield, Mass. The Massachusetts cases belong to the genera Catopteris and Ischypterus, with the exception of the Chicopee Falls specimens, which have been described under the name of *Acentrophorus chicopensis* Newb. The plants found all seem, according to Newberry, to be floated fragments which sunk and were buried. In Massachusetts they have been found at Turner's Falls, Sunderland and East Hampton, Montague and Mount Holyoke. The following species have been described by Newberry: *Schizoneura planicostata* Rogers, *Pachyphyllum simile* Newb., *Clathropteris platyphylla* Brong. (=*C. rectiusculus* Hitch.). From Mount Holyoke Hitchcock described *Tæniopteris* sp., and from Montague several remains of Voltzia?

The black fish beds of Turner's Falls and Sunderland lie just above the Deerfield trap sheet and the black plant beds of the Holyoke area lie above the Holyoke traps (Emerson).

JURASSIC (?), CRETACIC, EOCENE, NEOCENE.

1. GAY HEAD.

Route.—By rail, N. Y., N. H. & H. railroad, Taunton div. (Park square station) to New Bedford; thence by steamer to Cottage city, Martha's Vineyard. (Steamer touches at Wood's Holl and may be taken there.) Drive to Gay Head (roads not good for bicycle). Use topographic map as guide. At least three days should be allowed for this excursion.

The most recent section of the Gay Head strata is that given by Mr. J. B. Woodworth.[1] In descending order the following beds occur:

---

[1] J. B. Woodworth. Unconformity of Martha's Vineyard, etc. Geol. Soc. Amer., Bull., vol. 8, pp. 197-212, pl. 16.

Pleistocene:
  Moraine of Martha's Vineyard and Block island.
  Unconformity.
  Tisbury beds (not well exposed).
  Unconformity.
  Sankaty sands and gravels.
  Lower boulder bed.
  Unconformity.
Pliocene:
  Pliocene sands.
  Unconformity (inferred).
Miocene:
  Foraminiferal or Greensand beds.
  Unconformity.
  Osseous conglomerate.
  Unconformity.
Cretacic:
  Marine upper Cretacean.
  Unconformity (inferred).
  Non-marine plant-bearing beds (regarded by Marsh as Jurassic).

Jurassic (?) or lower Cretacean.

These are the lowest beds exposed at Gay Head. White in 1889 collected Cretacic plants in large numbers. The best were obtained "(1) from argillaceous concretions in the lignites and carbonaceous clays, (2) from the clay and clay-ironstone concretions in the reddish and gray clays, and (3) rarely from the concretions and the limonite matrix of the ferruginous conglomerates on the western escarpment."[1] The plants include cryptogams, conifers, monocotyledons, dicotyledons, and numerous fruits. The most important species are:

  Sphenopteris grevillioides Hr.
  Sequoia ambigua Hr.
  Andromeda parlatorii Hr.
  Myrsine borealis Hr.
  Liriodendron simplex Newb.
  Eucalyptus geinitzi Hr.
  Sapindus cf. *morrisoni* Lx.

[1] White, *loc. cit.*, p. 97.

These species (with the exception of the first) and a number of others from Gay Head are found in the middle Cretacean of other districts; to which, and more particularly to the horizon of the Amboy clays, these beds are referred. Professor Marsh, however, considers these beds to be the equivalent of the Potomac formation of Maryland; which he regards, from its vertebrate remains, as of Jurassic age. Total thickness probably less than one hundred and fifty feet (Woodworth).

Marine upper Cretacean.

These beds, exposed in the Indian hill district (described below), are probably represented at Gay Head, appearing stratigraphically beneath the Miocene beds (Woodworth). No fossils have been reported from the Gay Head section. Shaler considered the Indian hill beds as of middle Cretacean age.

Eocene (absent).

Neocene.

Miocene. From collections made by Woodworth and others, Dall identified the Greensand beds and underlying Osseous conglomerate as of Miocene age. In his paper will be found a list of species, which include eight vertebrates in addition to numerous unidentified remains of osseous fishes, and three crustaceans, *Archæoplax signifera, A.* ? sp. and *Balanus* (? *proteus*). Fourteen species of mollusca are reported from Gay Head, and three additional ones from Chilmark. The genera include Pecten, Yoldia, Nucula, Astarte, Crassatella, Cardium, Venus, Cytherea, Tellina, Macoma, Mya and others. The shells are represented mainly by internal, or in a few cases by external, casts.

The Osseous conglomerate, twelve to eighteen inches thick, is seen just north of the Devil's Den. It contains black chert pebbles bearing corals, crinoid stems, graptolites and shells; which indicate, according to Walcott, that they are derived from Silurian (Ordovician?) strata. The principal fossils of the conglomerate are the vertebræ, jaw-fragments, ribs, paddles, and head bones of cetaceans, and masses of lignite derived from the erosion of the Potomac series (Woodworth). The overlying Greensand varies up to ten feet in thickness. The lower beds are green, the upper rusty brown. The cast of *Macoma lyellii* are in the attitude of growth. "The crab *Archœoplax signifera* is mainly found in the lower portion of this stratum . . . " (Woodworth).

The Miocene is nowhere more than ten feet thick; and usually, by reason of erosion, it is much less. It rests unconformably on the Cretacic, the Eocene being absent.

Pliocene. From a series of fossils collected by Woodworth from the sands overlying the Greensand beds, Dall identified the following species:

Venericardia borealis Conr.
Astarte castanea Say.
Spirula polynyma Stm.
Corbicula densata Conr.
Macoma lyellii Dall?
Nucula shaleri Dall var.?
Purpura lapillus L.

Dall holds that "on the whole these specimens indicate a more recent fauna than the Miocene . . . and may perhaps be regarded as representing the Pliocene."[1]

2. INDIAN HILL.

(Cretacic.)

Professor Shaler has described fossiliferous Cretacic rocks from the north shore of Martha's Vineyard.[2] The locality is nearly south of Indian hill, in immediate proximity to the Martha's Vineyard esker, and a few hundred feet east of a ruined building known as Wood's schoolhouse. "The schoolhouse of the name has disappeared, for its foundations only remain; but the explorer can readily find his way to the spot by passing from the new schoolhouse on the Cedar Tree Neck road westwardly along the serpent kame, the only deposit of this nature on the island, until he passes a stone wall, a little to the west of which, in the roadway and on the bare ground thereabout, he may find an abundance of fragments of this peculiar sandstone."[3] The sandstone is of a reddish color, coarse, and abounds in quartz pebbles, and the fragments are angular. They are mingled with the till, and

---

[1] Dall, *loc. cit.*, p. 300.
[2] N. S. Shaler. Cretaceous fossils on Martha's Vineyard. Mus. Comp. Zoöl., Bull., vol. 16, no. 5.
[3] Shaler, *loc. cit.*, p. 90.

so numerous that many tons could be collected. These fragments are found over an area of less than an acre. Professor Shaler holds that the beds from which the fragments were derived " are in place at some little depth beneath the surface; within a few hundred feet of the locality where the Cretaceous waste now lies."[1]

The most abundant fossil in this rock is an Exogyra, different from any of the previously described species. Shaler holds that this form and the Camptonectes indicate that the deposits are of middle Cretacean or earlier age.

The following list is taken from Shaler's paper:
1. New genus? Cf. Myoconcha.
2. Plicatula or Ostrea. Cf. *P. instabile* Stol. and *O. lugubris* Conrad.
3. *Tellina (linearia)?*
4. Cardium?
5. Pteria.
6. Lucina?
7. *Turritella (nerina)?*
8. *Camptonectes burlingtonensis* Gabb.
9. *Camptonectes parvus* (?) Whitfield.
10. Chemnitzia.
11. Lucina.
12. Corithium.
13. Anomya?
14. Turritella.
15. Nuculana.
16. Ostrea or Exogyra?
17. Modiola.
18. Modiola?
19. Exogyra. Cf. *E. ostracina* Lam.

Another less prolific locality occurs, according to Professor Shaler, on the western shore of Lagoon pond, immediately west of Cottage city.

3. HIGHLAND LIGHT.

(Eocene erratics.)

Route.—By rail, N. Y., N. H. & H. railroad, Cape Cod div., Kneeland street station, to North Truro, 114 miles; fare $2.60. Thence walk or ride

[1] Shaler, Geol. Martha's Vineyard. U. S. Geol. Surv., 7th Ann. rep., p. 326; 1888.

to Highland light, two miles. Fair bicycle roads. *By boat*, from Commercial wharf, Atlantic avenue, to Provincetown; fare, round trip, $1.50 (see daily papers). Thence drive to Highland light, seven miles. Roads sandy and hilly. Arrangements can be made for a carriage or barge for the whole trip.

About half a mile south of Highland light the cliff has a height of about one hundred and fifty feet, and consists of sand and gravel, the pebbles being of all sizes up to a foot in diameter, mostly rounded, but a part of them angular. Among the pebbles and boulders fallen on the beach are found frequently pieces of rock containing shells and fragments of shells, as well as vestiges of lignite and other organic remains. Professor Crosby has identified the following Eocene species from collections made by Mr. Upham:

Venericardia planicosta, *V. parva?*, *V. alticosta?*, Ostrea, three species, including *O. divaricata* (?) or young of *O. sellæformis*, and *O. virginiana*, *Anomia tellinoides?*, *Plicatula filamentosa*, *Camptonectes calvatus*, *Axinoma staminea*, *Striarca centenaria*, *Cardium* sp., *Yoldia* sp., *Corbula* sp., *Natica*, sp., other gastropods, spines of Cidaris, and a coral resembling Galaxea. Crosby thinks that these fossiliferous pebbles indicate that Eocene beds are somewhere in place under Massachusetts bay. Similar fragments were obtained by Verrill from the Grand Bank and George's Bank.

POST-PLIOCENE FOSSILS.

1. SANKATY HEAD, NANTUCKET.

Route.— *By steamer*, from New Bedford, Wood's Holl or Cottage city (see Gay Head) to Nantucket, thence by Nantucket railway to Siasconsett. Short drive or walk to Sankaty Head.

Four different sections of the beds at Sankaty Head have been published. These are: in 1847 by Messrs. Desor and Cabot, in 1874 by Mr. S. H. Scudder, in 1889 by Prof. N. S. Shaler, and in 1895 by Mr. Frederick J. H. Merrill. These sections differ somewhat among themselves, due probably in a large degree to the erosional changes which have occurred during the intervals between successive observations.[1]

---

[1] A complete résumé of the earlier papers is given by Shaler in his Geology of Nantucket, Bull. 53, U. S. Geol. Survey, which should be taken into the field.

The section, according to Merrill, is as follows:[1]

		ft.	in.
1.	Fine dark drifted sand,	3	
2.	Yellow sandy drift ferruginous at the bottom, and containing pebbles,	5	
3.	Coarse gray stratified sand, with particles of green sand,	40	
4.	Fine white clayey sand, with ferruginous streaks, and very minute particles of green sand,	10	
5.	Fragment bed,	1	
6.	Upper shell bed,		8
7.	Clayey ferruginous sand,		4
8.	Serpula sand,	1	3
9.	Lower shell bed,		9
10.	Red sand, with fragments of blue clay,	1	
11.	White sand of varying quality and size,	4	
	Concealed by turf and beach sand,	24	
	Total,	90	

According to Scudder, below the bed of white sand (No. 11) occur four feet of coarse gravel and sand, and below that the basal beds of light brown sandy clay.

In the oyster bed (lower shell bed) the shells commonly lie in their natural position, with both valves together (Desor), but Merrill states that they often lie in all positions, with valves separated. The most abundant species are *Ostrea virginiana, Venus mercenaria, Modiola hamatus, Cummingia tellinoides, Arca transversa* and *Urosalpinx cinerea.* "The assemblage of species is similar to that now living in the protected bays of southern New England, at a depth of three to five fathoms" (Verrill). The Serpula bed consists mainly of convoluted masses of *S. dianthus*, which still occurs in abundance on the southern coast of New England, in sheltered bays and harbors. At the bottom, according to Merrill, the bed consists of detached masses of the Serpula, closely packed together; while at the top the tubes are very much comminuted.

The earlier observers had found scarcely any other species than *Serpula dianthus* in the Serpula bed; but Merrill gives a list

[1] Merrill, *loc. cit.*, p. 11.

of sixteen species, in addition to the bryozoan *Hippothoa variabilis* which encrusts the worm tubes. The layer of clayey ferruginous sand overlying the Serpula bed contains *Ostrea virginiana* and small specimens of *Argina pexata*. "The upper shell bed consists of coarse beach sand, with pebbles, slightly intermixed with clayey matter" (Merrill). The "fragment bed" of Merrill, or the uppermost fossiliferous stratum (included by Scudder in the upper shell bed), "consists of white quartz sand and pebbles, with a great abundance of comminuted shells." It differs from the upper shell bed proper in the condition of the fossils and the entire absence of ferruginous matter.

The strata above the Serpula bed are especially rich in northern species, such as *Buccinum undatum*, *Ceronia arctata*, *Astarte castanea*, *Cyclocardia borealis*, *Mya truncata*, *Balanus porcatus*, etc., which are common. Many of these are now found on the outer shores of Cape Cod.

The total number of species from the three beds, as given by Verrill, is sixty,—most of them mollusca. Thirty-seven species are found in the lower and thirty-six in the upper bed. Only thirteen species are common to the two.

Merrill has made some additions to this list, and also has shown a greater vertical distribution for some of the species. He regards these deposits as transported material, thus differing from Verrill; who, deducing from the observations of others, regarded the oyster and Serpula beds as deposited in a shallow bay protected from the outer waters by a barrier, and considered that the upper bed was deposited after the breaking of the bar and the ingress of the Atlantic waves.

2. WINTHROP GREAT HEAD.

Route.—*By train*, Boston, Revere Beach and Lynn railroad (narrow gauge), Atlantic avenue station, to Winthrop beach, then walk to cliff.

By electrics, from Scollay square, Beachmont car to Beachmont, then follow shore past Grover's cliff, to Winthrop Great Head cliff. Or take East Boston car *via* ferry, changing to Winthrop car in East Boston.

This is the original locality from which Stimpson obtained his recent fossils in 1854. His list included: *Balanus rugosus* (= *B. crenatus*), *Mya arenaria*, *Solen ensis* (= *Ensatella americana*) *Mactra solidissima*, *Venus mercenaria*, *Astarte sulcata* (= *A. undata*), *A. castanea*, *Cardita borealis* (= *Cyclocardia borealis*),

Mytilus edulis, Modiola modiolus, Ostrea borealis (*O. Virginiana*), *Fusus decemcostatus* (= *Chrysodomus decemcostatus*), *Buccinum plicosum* (= *Urosalpinx cinerea*), *B. trivittatum* (=*Tritia trivittatum*). All these were found alive by dredging within a mile of the locality. The outcrop of the shell beds is fifty or sixty feet above the sea level, and the shells are mostly broken and often finely comminuted. Numerous specimens have been collected since from this and other drift sections in Boston harbor by Upham, Dodge, Herman, Crosby and others, who brought the number of species up to thirty-four previous to 1894. In that year, Prof. W. O. Crosby and Miss Hetty O. Ballard published a joint paper on the drift fossils, in which were embodied the results of a careful study of collections made by them with the assistance of several students of the Institute of Technology As a result of these studies, the number of species was brought up to fifty-five, and the number of localities to twenty-four. The most important of these localities, with the number of species found in each are as follows:

Great Head	32
Grover's cliff	32
East End Peddock's island	29
West " " "	27
Deer island	21
Moon island	16
Telegraph hill	19
Point Allerton	9
Great Brewster island	8

3. GLOUCESTER.

Route.—*By rail*, Boston and Maine railroad, North Union station, to Gloucester (see p. 15). About three hundred yards south of the Pavilion hotel (Gloucester), and between that building and a small earthwork battery known as Stage Fort, is an exposure in a cliff about twenty-five feet high, in which the fossils were found. Professor Shaler states that this exposure is again accessible.

This section was studied by Shaler in 1868, who reported the cliff as consisting at the base of "close grained, much indurated sands, which have acquired compactness by their own weight without the influence of other metamorphic action." About ten feet above high-water mark are some thin layers of a more clayey texture than is the main mass; and in this, and a little distance above and below, all the fossils were found. The same sands as below succeed this bed, capped by ordinary semi-stratified pebbly drift.

The following partial list of fossils is given by Shaler:

Leda. Two specimens.
Modiola discrepans Say. Several.
Mya truncata, Linn.? "
Mesodesma arctata? Very doubtful.
Nucula sapotilla ?
Panopea arctica Gould.?
Saxicava distorta Say.
Five or six specimens of lamellibranchiates, not identified.
Crustacean remains plentiful but very fragmentary.

4. BOSTON.

(Undisturbed Recent fossiliferous beds.)

Post-Pliocene fossils of very recent date have been obtained at three localities in Boston; at each of which they were, unlike the drift shells, taken from the beds in which they were originally deposited. The fossils were brought to light by excavating; and the localities are now covered over so that at present no more specimens can be obtained. The fossils were listed in several papers by Upham, and the specimens may be seen now in the collections of the Boston Society of Natural History, or the Institute of Technology. Some of the rarer species are in the possession of one of the collectors, Mr. Warren Herman.

A. Valley of Muddy river, Brookline district.—Here an excavation for a sewer exposed a fossiliferous clayey stratum, near the present level of low tide. It is underlain by stratified clays, and overlain by about one foot of peat, succeeded by five to twelve feet of muddy alluvium. Thirteen species were found, most of them molluscs now inhabiting this coast.

B. North bank of Charles river at Cambridge end of Harvard bridge.—Here all the land lying between the river and the railroad was built up from the dredgings taken from the bottom of the river. Below the river mud, sands were met containing twelve species of fossils. These were brought up by the dredge in vast numbers. The most abundant species were *Ostrea virginiana*, *Mya arenaria*, *Venus mercenaria* and *Pecten irradians*; of these *Mya arenaria* is the only species now occurring in abundance north of Cape Cod, the others being found only in isolated " colonies." The size of the *Mya* and oyster shells was found to be enormous—a valve of the latter, recorded by Upham, having a thickness of one and one-half inches. A length of ten inches is not uncommon among these shells. They are buried again by the thousands just beneath the surface of this new-made land, where any future digging will expose them.

C. *City point, South Boston* (reached by electric cars from Park square).—In the dredgings made to fill up the land for the Marine park, many shells were brought up in the mud in which they had been buried recently. Twenty-one species were identified and recorded by Upham.

The total number of species found in these three localities was originally made twenty-five by Upham. The collections chiefly from South Boston, of Mr. Warren Herman, have furnished twenty-six more species, making a total of fifty-one species so far recorded. Twenty-four of these are distinctly southern at the present time, reaching their northward limit at Cape Cod or Massachusetts bay, though some extend in isolated colonies farther north. This indicates a period of warmer waters preceding the present and succeeding the glacial episode. (For a complete discussion see Upham '93.)

LITERATURE.

(Palæontology of eastern Massachusetts.)

(NOTE:— While this bibliography aims to take account of all the important contributions to the palæontology of eastern Massachusetts, it must necessarily be imperfect, on account of the very scattered distribution of the articles. Several minor papers, not dealing directly with the palæontological side of Massachusetts geology, purposely have been omitted.) Consult also Hitchcock's Geology of Massachusetts.

The most important papers are starred.

1. CAMBRIC.

1834. *Green, Jacob.* Descriptions of some new North American Trilobites. (Am. Journ. Sci., vol. 25, pp. 334–337.) Describes *Paradoxides harlani*.

1856. *Rogers, W. B.* Notes on Paradoxides from Braintree. (Bost. Soc. Nat. Hist., Proc., vol. 6, pp. 27-29, 40-44.) Announcement of first discovery of trilobites *in situ* in Braintree, and history of first specimen.

1856. *Stodder, Charles.* Note on a specimen of *Paradoxides harlani* from Braintree. (Bost. Soc. Nat. Hist., Proc., vol. 6, p. 369.) Describes a well-preserved specimen.

1856. *Rogers, W. B.* Discovery of Palæozoic fossils in eastern Massachusetts. (Am. Journ. Sci., 2d ser., vol. 22, p. 296-298.) Announces discovery of Paradoxides at Braintree.

1856. *Rogers, W. B.* Proofs of the Protozoic age of some of the altered rocks of eastern Massachusetts, from fossils recently discovered. (Am. Acad. Sci., Proc., vol. 3, pp. 315-318.) Announcement of discovery of Paradoxides at Braintree, and discussion of its bearing on the classification of our rocks.

1858. *Lea, Isaac.* On the Trilobite formation at Braintree, Mass. (Phil. Acad. Sci., Proc., vol. 9, p. 205.) Description of quarry and occurrence of fossils.

1859. *Jackson, C. T.* Note on *Paradoxides harlani* and related species. (Bost. Soc. Nat. Hist., Proc., vol. 7, p. 54.) Compares *P. harlani* with a species from St. Mary's bay, N. F. (p. 75) and gives measurements of two specimens.

1859. *Rogers, W. B.* Note on Paradoxides. (Bost. Soc. Nat. Hist., Proc., vol. 7, p. 86. Compare *P. harlani* and *P. noviapertus*.

1860. *Barrande, J.* Note sur la faune primordiale. (Bull. de la Soc. Geol. de France, tome 17, pp. 542-554.) Comparison of *P. bennetti* and *P. spinosus* (*P. harlani*).

1860. *Barrande, J.* On the Primordial fauna and the Taconic system. (Bost. Soc. Nat. Hist., Proc., vol. 7, p. 369.) *P. harlani* identical with *P. spinosus* of Bohemia.

1860. *Ordway, Albert.* Notes on *Paradoxides harlani*. (Bost. Soc. Nat. Hist., Proc., vol. 7, p. 427.) Comparison of *P. harlani* and *P. spinosus* proves identity of each.

1861. *Ordway, Albert.* On the supposed identity of *Paradoxides harlani* Green, with the *Paradoxides spinosus* Boeck. (Bost. Soc. Nat. Hist., Proc., vol. 8, pp. 1-5, figs. 1, 2.) Compares the two species and gives figures of each. The species are distinct.

1861. *Jackson, C. T.* Note on Paradoxides. (Bost. Soc. Nat. Hist., Proc., vol. 8, p. 58.) *P. harlani*, *P. bennetti* and *P. spinosus* closely related.

1861. *Rogers, W. B.* On fossiliferous pebbles of Potsdam rocks in Carboniferous conglomerate north of Fall River, Mass. (Bost. Soc. Nat. Hist., Proc., vol. 7, pp. 389-391.) Quartzite pebbles contain Lingula, resembling *L. prima* and *L. antiqua*. (Specimens now in Bost. Soc. Nat. Hist. collection.)

1861. *Marcou, Jules.* On the black slate from Braintree, Mass., containing Paradoxides, etc. (Bost. Soc. Nat. Hist., Proc., vol. 7, pp. 357, 358.) Compares Braintree beds to those of Newfoundland

1884. *Whitfield, R. P.* Notice of some new species of Primordial fossils, etc. (Am. Mus. Nat. Hist., Bull., vol. 1, p. 147.) Describes *Arionellus* (=*Agraulus*) *quadrangularis*, from Braintree.

1884. **Walcott, C. D.* On the Cambrian faunas of North America. (U. S. Geol. Surv., Bull. 10, "Fauna of the Braintree argillites," pp. 43–49, pls. 7-9.) Describes locality and *Hyolithes shaleri, Paradoxides harlani, Ptychoparia rogersi,* and *Agraulus quadrangularis*.

1888. **Shaler, N. S.* and *Foerste, A. F.* Preliminary description of North Attleborough fossils. (Mus. Comp. Zool., Bull., vol. 16, pp. 77-91, pls. 1 and 2.) Describes twenty-one species from lower Cambrian.

1889. *Foerste, Aug. F.* Palæoutological horizon of the limestone at Nahant. (Bost. Soc. Nat. Hist., Proc., vol. 24, pp. 261-263.) Announces discovery of *Hyolithes inæquilateralis* Foerste (= *H. communis* var. *emmonsi* Walcott) in limestone of Nahant, which is Olenellus Cambrian.

1890. *Sears, J. H.* The stratified rocks of Essex county. (Essex Inst., Bull., vol. 22, pp. 31–47.) Fossiliferous strata at Nahant, Rowley, Topsfield and Jeffrey's ledge. All are lower Cambrian.

1893. *Woodworth, J. B.* Note on the occurrence of erratic Cambrian fossils in the Neocene gravels of the island of Martha's Vineyard. (Am. Geol., vol. 9, pp. 243-247.) Olenellus Cambrian fossils in pebbles derived from a conglomerate bed.

1893. *Walcott, C. D.* Note on lower Cambrian fossils from Cohasset, Mass. (Biol. Soc. Wash., Proc., vol. 7, p. 155.) Mentions occurrence of *Hyolithes communis* and *Straparollina remota* in boulder from Cohasset.

1894. *Woodworth, J. B.* On traces of a fauna in the Cambridge slates. (Bost. Soc. Nat. Hist., Proc., vol. 26, pp. 125-126.) Trails similar to those made by modern isopods, obtained from Malden, Mystic river, Clarendon hills. Monocraterion and pits like the borings known as Arenicolites occur. Age not known.

2. DEVONIC.

1851. *Hitchcock, E.* On the geological age of the clay slates of the Connecticut valley in Massachusetts and Vermont. (Am. Assoc. Adv. Sci., Proc., vol. 6, pp. 299-300.) A fossiliferous limestone discovered at Bernardston, Mass., containing crinoid stems which by Hall are referred to the lower Devonian (originally described in 1835 in 2d edit. Geol. Mass.).

1861. *Hitchcock, C. E.* (Rep. Geol. Vermont, p. 447.) Detailed description of the locality with correlation.

1873. *Dana, J. D.* On rocks of the Helderberg era, in the valley of the Connecticut, etc. (Am. Journ. Sci., 3d ser., vol. 6, pp. 339-352). Discusses age of rocks and gives map of locality.

1877. *Dana, J. D.* Note on the Helderberg formation of Bernardston, Mass., etc. (Am. Journ. Sci., 3d ser., vol. 14, pp. 379-387.) Discusses superposition and age of beds.
1877. *Hitchcock, C. H.* Note upon the Connecticut valley Helderberg. (Am. Journ. Sci., 3d ser., vol. 13, p. 313.)
1883. *Whitfield, R. P.* Observations on the fossils of the metamorphic rocks of Bernardston, Mass. (Am. Journ. Sci., 3d ser., vol. 25, pp. 368-369) with a note by Dana on the fossiliferous shale. Mentions occurrence of over seven recognizable fossils in the shaly quartzite and several species in the limestone below. Regards limestone as middle Silurian and quartzite beds as middle Devonian.
1890. *Emerson, B. K.* A description of the Bernardston series of metamorphic upper Devonian rocks. (Am. Journ. Sci., 3d ser., vol. 40, pp. 263-275, 363-374.) Note by J. D. Dana. Elaborate discussion of the beds and their relations. Most abundant and characteristic fossils are Chemung with Hamilton forms.

3. CARBONIC.

1844. *Teschemacher.* Ferns from coal of Mansfield, Mass. (Bost. Soc. Nat. Hist., Proc., vol. 1, p. 62.) Mentions occurrence of seven species of ferns in fine state of preservation. Note by C. T. Jackson.
1851. *Jackson, C. T.* Note on a fossil Calamite from Bridgewater, Mass. (Bost. Soc. Nat. Hist., Proc., vol. 3, p. 223.) Brief description of specimen.
1880. *Crosby, W. O.* and *Barton, G. H.* Extension of Carboniferous formation in Massachusetts. (Am. Journ. Sci., 3d ser., vol. 20, pp. 416-420.) Fossil Calamites and Sigillaria (?) found at Rockdale, Mass. Represented by hollow moulds in rock.
1881. *Barton, G. H.* Norfolk County basin, Mass. (Science Observer, vol. 3, pp. 41-42.)
1885. *Perry, Joseph.* Note on a fossil coal plant found at the graphite deposit in mica-schist at Worcester, Mass. (Am. Journ. Sci., 3d ser., vol. 29, pp. 157-158.) Two specimens of *Lepidodendron acuminatum* from mica-schist surrounding granite knoll.
1887. *Kemp, J. F.* Fossil plants and rock specimens from Worcester, Mass. (N. Y. Acad. Sci., Trans., vol. 4, pp. 75-76.) Notice of Perry's plants from Worcester.
1894. *Woodworth, J. B.* Carboniferous fossils in the Norfolk County basin. (Am. Journ. Sci., 3d ser., vol. 48, pp. 145-148.) At Canton Junction occur Calamites, Sigillaria and fern stems.
1896. *Fuller, Myron L.* A new occurrence of Carboniferous fossils in the Narragansett basin. (Bos. Soc. Nat. Hist., Proc., vol. 27, pp. 195-199.) Occurrence of Calamites and Sigillaria (?) at Brockton, Mass.

4. Triassic.

1836. *Hitchcock, E.* Ornithichnology. Description of the footmarks of birds (Ornithichnites) on New Red sandstone in Massachusetts. (Am. Journ. Sci., vol. 29, pp. 307-340, 2 plates.) Describes seven species of "Ornithichnites" considered bird tracks.

1843. *Lyell, Charles.* On the fossil footprints of birds and impressions of raindrops in the valley of the Connecticut. (Am. Journ. Sci., vol. 45, pp. 394-397.) Account of visit to principal quarries showing impressions.

1843. *Hitchcock, E.* Description of several species of fossil plants from the New Red sandstone formation of Connecticut and Massachusetts. (Am. Assoc. Geol. and Nat., Trans., pp. 294-296, pp. 12-13.) Found Tæniopteris in boulder at Mount Holyoke and several small plant remains of Voltzia (?) at Montague, Mass.

1844. *Deane, James.* On the fossil footmarks at Turner's Falls, Mass. (Am. Jour. Sci., vol. 46, pp. 73-77, 2 plates). Describes Turner's Falls, and tracks occurring there. Regards them as bird tracks.

1847. *Deane, James.* Illustrations of fossil footmarks. (Boston Journ. Nat. Hist., vol. 5, pp. 277-284.) Discusses footmarks, which could be made only by birds. Discusses distribution and character of the birds.

1854. *Hitchcock, E.* On the fossil footmarks, sandstones, and traps of the Connecticut valley. (Bost. Soc. Nat. Hist., Proc., vol. 4, pp. 378-379.) Gives localities for footprints. Discusses age of rocks.

1855. *Hitchcock, E. jr.* Description of a new species of Clathopteris in the Connecticut valley sandstone. (Am. Journ. Sci., 2d ser., vol. 20, pp. 22-25.) Describes *C. rectiusculus* Hitch. from Mount Tom, Easthampton, Mass. Genus a Jurassic one.

1858. * *Hitchcock, E.* Ichnology of New England. A report on the sandstones of the Connecticut valley, especially its fossil footmarks, vii + 220 pages, 60 plates, 4to. Elaborate discussion, description and illustration of specimens. Refers beds bearing footprints, fishes and ferns to lower Jurassic. Gives thirty-eight localities for footprints.

1861. *Deane, James.* Ichnographs from the sandstone of the Connecticut river. Boston 1861. (Not seen.)

1865. * *Hitchcock, E.* Supplement to the Ichnology of New England. Boston 1865. Describes thirty-seven new species. Appendix with description of bones of *Megadactylus polyzelus* Hitch. and catalogue of specimens in Amherst college.

1867. *Shepard, Charles.* On the supposed tadpole nests or imprints made by the *Batrachoides nidificans* Hitch. in the red shale of the New Red sandstone of South Hadley, Mass. (Am. Journ. Sci., 2d ser., vol. 43, pp. 99-104.) Describes slates with peculiar impressions regarded formerly as of organic origin, and holds them to be the results of cross-ripples.

1888. *Newberry, J. S. Fossil fishes and fossil plants of the Triassic rocks of New Jersey and the Connecticut valley. (U. S. Geol. Surv., Mon. 14.) Description of species.

1891. Davis, W. M. Two belts of fossiliferous black shale in the Triassic formation in Connecticut. Discussion. (Geol. Soc. Am., Bull., vol. 2 [pp. 415-424,] p. 430.) The fish and plant beds of the Massachusetts Triassic should be found on the back of Mount Tom-Holyoke range, confirmed by Bear's Hole locality, a mile or two north of Westfield river.

1891. Emerson, B. K. Stratigraphic position of fossil-bearing beds in the Newark formation in Massachusetts. (Geol. Soc. Am., Bull., vol. 2, p. 430 (4 lines). Band of Black Shale occupies horizon above Holyoke traps. Furnished only plant remains. Sunderland and Turner's Falls fish beds just above Deerfield trap sheet.

1892. Mitivras, M. M. Footprints from Connecticut valley. (Am. Assoc. Adv. Sci., Proc., vol. 40, p. 286, abstract.)

5. JURASSIC.

1896. *Marsh, O. C. The Jurassic formation on the Atlantic coast. (Am. Journ. Sci., 4th ser., vol. 2, pp. 433-447.) Gives evidence for the belief that the plant-bearing beds of Martha's Vineyard are Jurassic.

6. CRETACIC.

1860. Stimpson, William. Cretaceous strata at Gay Head, Mass. (Am. Jour. Sci., 2d ser., vol. 29, p. 145 ($\frac{1}{4}$ p.).) Gives list of fossils collected, and regards them as Cretacic. Many of these were from the Miocene beds — bones, vertebræ, shark's teeth, crustacea, twelve species bivalves, one univalve, leaves, seeds, etc.

1890. *White, D. Cretaceous plants from Martha's Vineyard. (Am. Jour. Sci., 3d ser., vol. 39, pp. 93-101, pl. II.) Occurrence of plants with descriptions.

1890. White, D., Newberry, J. S., Ward, L. F.; Merrill, F. J. H. Cretaceous plants from Martha's Vineyard. (Geol. Soc. Am., Bull., vol. 1, pp. 554-556.) Discussion of White's paper. General consensus of opinion as to Cretacic age of plant beds.

1890. *Shaler, N. S. On the occurrence of fossils of the Cretaceous age on the island of Martha's Vineyard. (Mus. Comp. Zool., Bull., vol. 16, pp. 89-97, pls. 1, 2.) Describes nineteen species from sandy beds near Indian hill. Regards deposits as middle Cretacean.

1893. Uhler, P. R. Gay Head. (Science, vol. 20, pp. 176, 177.) Description of Gay Head beds.

1893. White, David. The Cretaceous at Gay Head, Martha's Vineyard. (Science, vol. 20. pp. 332, 333.) Criticism of Uhler's succession and correlation of strata.

1893. *Uhler, P. R.* Observations on the Cretaceous at Gay Head. (Science, vol. 20, pp. 373, 374.) Reply to David White.
1896. *Hollick, Arthur.* Martha's Vineyard Cretaceous plants. (Geol. Soc. Am., Bull., vol. 7, pp. 12–14. Abstract.) Gives list of prominent species, correlates plant-bearing beds with Amboy clays of New Jersey.

7. EOCENE.

1881. *Crosby, W. O.* On the occurrence of fossiliferous boulders in the drift of Truro, Cape Cod. (Bost. Soc. Nat. Hist., Proc., vol. 20, pp. 136–140.) Boulders found by Upham south of Highland light contain Eocene fossils. Probably derived from Eocene deposit in Boston harbor.

8 NEOCENE.

1844. *Lyell, Charles.* Tertiary of Martha's Vineyard. (Am. Journ. Sci., 1st ser., vol. 46, pp. 318–320.) Collected canine tooth of seal, skull of walrus, vertebræ of cetacea, shark's teeth, two species of crustacea, two Tellinas, a Cytherea and three species of Mya. Concludes that strata are Miocene.
1894. *Dall, William H.* Notes on the Miocene and Pliocene of Gay Head, Martha's Vineyard, Mass. etc. (Am. Journ. Sci., 3d ser., vol. 48, pp. 296–301.) Gives a list of Miocene fossils (see text). Describes *Nucula shaleri* and *Macoma lyellii*. Gives list of Pliocene fossils.

9. POST-PLIOCENE.

A. Nantucket.

1849. *Desor, E.* and *Cabot, Edward C.* On the Tertiary and more recent deposits in the island of Nantucket. (Quart. Journ. Geol. Soc., vol. 5, pp. 340–342.) Section and list of fossils from Sankaty Head.
1849. *Desor, E.* On a deposit of drift shells in the cliffs of Sancati, Island of Nantucket. (Am. Assoc. Adv. Sci., Proc., vol. 1, pp. 100–101.) List of species with discussion of beds.
1851. *Desor, E.* Drift fossils from Nantucket. (Bost. Soc. Nat. Hist., Proc., vol. 3, pp. 79–80.) General account.
1875. *Scudder, Sam'l H.* Note on the post-Pliocene strata of Sankoty Head. (Am. Journ. Sci., 3d ser., vol. 10, pp. 365–368.) Gives detailed section.
1875. *Verrill, A. E.* On the post-Pliocene fossils of Sankoty Head, Nantucket Island, etc. (Am. Journ. Sci., 3d ser., vol. 10, pp. 364–375.) Discusses deposits and fossils and gives complete list, sixty species in all. Gives present habitat of each species. Discusses origin of deposit.

1877. *Scudder, Sam'l H.* Post-Pliocene fossils near Sankoty Head, Nantucket. (Bost. Soc. Nat. Hist., Proc., vol. 18, pp. 182-185.) Gives résumé of two preceding papers.
1889. *Shaler, N. S.* The geology of Nantucket. (U. S. Geol. Surv., Bull. 53.) Complete discussion with résumé of previous papers.
1895. *Hollick, Arthur.* Geological notes. Long Island and Nantucket. (N. Y. Acad. Sci., Trans., vol. 15, pp. 6-10.) Present condition of cliff very unsatisfactory for collecting fossils, seventeen species collected. Two species are new. *Panopea* sp. and *Mesodesma jauresi.* Also a specimen of silicified palm wood.
1895. *Merrill, Frederick J. H.* Post-Pliocene deposits of Sankoty Head. (N. Y. Acad. Sci., Trans., vol. 15, pp. 10-16.) Gives section and lists of species found, adding to Verrill's list, and extending vertical range of species.

B. Drift fossils.

1854. *Stimpson, William.* A list of fossils found in the post-Pliocene deposits in Chelsea, Mass. (Bost. Soc. Nat. Hist., Proc., vol. 4, pp. 9-10.) List of species from Great Head, Winthrop.
1868. *Shaler, N. S.* Notes on the position and character of some glacial beds containing fossils at Gloucester, Mass. (Bost. Soc. Nat. Hist., Proc., vol. 11, pp. 27-30.) Gives list of species found in section of glacial clays at Gloucester.
1869. *Niles, W. H.* Recent shells at great depth below fort Warren. (Bost. Soc. Nat. Hist., Proc., vol. 12, pp. 244-364.) Recent finding of shells of *Lunatia heros, Venus mercenaria* and *Cardita borealis* one hundred feet below surface at fort Warren.
1888. *Dodge, W. W.* Some localities of post-Tertiary and Tertiary fossils in Massachusetts. (Am. Journ. Sci., 3d ser., vol. 36, pp. 56-57.) Found several fossils in section at Winthrop Great Head.
1889. *Upham, Warren.* Marine shells and fragments of shells in the till near Boston. (Bost. Soc. Nat. Hist., Proc., vol. 24, pp. 127-141, and Am. Journ. Sci., vol. 37, pp. 359-372.) Gives list of twenty-one species and detailed discussion of origin.
1893. *Crosby, W. O.* Geology of the Boston Basin, vol. 1, pt. 1. (Bost. Soc. Nat. Hist., Occ. Papers, vol. 4.) Eleven species of fossils collected from Telegraph hill, Point Allerton, Great hill and Strawberry hill, Nantasket.
1894. *Dodge, R. E.* Additional species of Pleistocene fossils from Winthrop, Mass. (Am. Journ. Sci., 3d ser., vol. 47, pp. 100-104.) Makes addition of a few species to list from Winthrop Head.
1894. *Upham, Warren.* Marine shell fragments in drumlins near Boston, Mass. (Am. Journ. Sci., 3d ser., vol. 47, pp. 238-239.) Remarks on papers by Dodge and Crosby and gives additional notes.

1894. *Crosby, W. O.* and *Ballard, H. O.* Distribution and probable age of the fossil shells in the drumlins of the Boston Basin. (Am. Journ. Sci., 3d ser., vol. 48, pp. 486-496.) Fifty-five species of fossils, from twenty-four localities. Detailed discussion of occurrence and source of fossils.

C. Undisturbed deposits.

1865. *Hitchcock, C. H.* Impressions (chiefly tracks) on alluvial clay in Hadley, Mass. (Am. Journ. Sci., 2d ser., vol. 19, pp. 391-393.) Impressions of thirteen kinds of animals (man, four birds, two quadrupeds, one batrachian, snails, annelids, two or three of doubtful character) on clay beneath twenty feet of alluvial sand with ferruginous concretions. Compares with Triassic tracks.

1868. *Stodder, Charles.* On a recent gathering of diatomaceous mud from Pleasant beach, Cohasset. (Bost. Soc. Nat. Hist., Proc., vol. 11, p. 132-134.) Many species of diatoms collected from mud of marsh directly in the rear of the Minot house.

1893. *Upham, Warren.* Recent fossils of the harbor and Back bay, Boston. (Am. Journ. Sci., 3d ser., vol. 43, pp. 201-209, and Bost. Soc. Nat. Hist., Proc., vol. 25, pp. 305-316, this latter with additional note.) List of twenty-five species from Brookline, Charles river and South Boston; to which are added from South Boston twenty-six species, making total of fifty-one, twenty-four of which are southern.

PETROGRAPHY.

Prof. J. E. Wolff.

Among the petrographical features in the vicinity of Boston may be mentioned the Quincy granite area on the south, with numerous quarries in the dark gray hornblende granitite. The contact with the slates is well shown in those on the northeast. The Blue Hills are a complex of eruptive rocks comprising various forms of granite-porphyry and quartz-porphyry (aporhyolite), collected between Rattlesnake hill and Wampatuck.

The Middlesex Fells to the north of Boston include granite, granite-porphyry, quartz-porphyry, tuffs and trap dikes. The complex of igneous rocks of Essex county from Cape Ann to Lynn is of great interest. Beginning with the granitites quarried at Cape Ann, syenite, elæolite syenite, diorite, essexite and numerous trap dikes are met along the coast. At Marblehead Neck is the original locality for Bostonite, and fine felsite-breccias are found, and granite intrusive into schist. Along the Clifton shore granite, diorite, pegmatite and trap dikes exist in very interesting relations.

The famous Medford-Somerville coarse diabase, so often described for its weathering, is well shown at Granite street, Somerville, and in the old quarry in the west side of Pine hill, Medford.

The dikes of the Cambridge-Somerville slates (Kidder avenue, etc.) are of interest. The various amygdaloids (Brighton, Hough's Neck, etc.) have been the subject of several papers.

PHOTOGRAPHING AND COLLECTING.

J. Edmund Woodman.

There are few regions in which such opportunities are offered for collecting representative specimens in variety, and for photographing forms, as in the vicinity of Boston. We are especially rich in petrographic material illustrating rare rocks or structural varieties and development, interesting to the petrographer more than to the general geologist. It is impossible without undue expansion of space to enumerate these opportunities, with the localities where they are offered. The following list, however, made for the use of one of the classes of the Summer School of Harvard University, will give some idea of the richness of the region. The excursions upon the basis of which this list was made are most of those included in this guide. Objects unmarked can be both photographed and collected; those marked *can only be photographed.

*Monaduocks; veins; *roches moutonnées; *marine-benches; *eroded igneous sea-cliffs, with characteristic outlines; flow structure in lavas; aporhyolites; *swinging sand-spits; *partially submerged and eroded drumlins; *swamping in quiet areas; marine pebbles; marine sand of different varieties; breccias; weathered diabase in all stages; concentric weathering of diabase; *talus cones of weathered material; *glacial striation; *river gorge; *pot-holes; *glacial interference with drainage; *cross-section and outline of drumlins; glacial till; *dike chasms; *till and sedimentary rocks in process of erosion by waves; *stacks; *natural bridges; *complicated faulting and intersections of dikes; igneous rocks — granite, diabase, diorite, quartz porphyry, etc.; *pocket beaches; *wall beaches; algæ protecting rocks; barnacles protecting rocks; *ripple-marks; *rill-marks, often with deltas; *wave-marks; *trails of organisms; porphyritic structure; *spouting horn; *cross-section of eskers; esker gravel of various sizes;

*cross-bedding in sand-plains; *fore-set and top-set beds; *lobes of sand-plains; glacial sand; *ice-contact slopes; *kettle holes; *kames; kame gravel; *dunes; eolian sand; *off-shore bars; metamorphic rocks and minerals; joint-blocks of many varieties; *sills; *folds; *joint chasms (marine); *joints in igneous rocks; *boulder moraines; sedimentary rocks — slate, shale, sandstone, quartzite, arkose, compound conglomerate; *waves in their action against the shore; relation of cleavage to bedding; amygdaloids; *moraine over stratified drift; *glacial cone (moraine); small faults; *intraglacial swamps; fossil rain-prints; *cross-bedding in old sediments; *comparison of lines of wave-wear and ice-wear.

II.

ZOOLOGY: MARINE INVERTEBRATES.

Amadeus W. Grabau.

The seashore is reached readily from Boston by train, electric cars, or bicycle. Most of the places described can be visited in half a day, although a whole day will be found none too long for those who wish to see the varied marine fauna, characteristic of the different localities. All of the localities should be visited at low tide, the time of which can be figured out from the times of the morning and evening high tide given in the daily papers. Extreme low tide (announced in the Old Farmer's and other almanacs) is best for all the localities, though the beaches furnish interesting material even at high water.

The Littoral and Laminarian zones are the only ones readily accessible, unless one is equipped with a dredge for obtaining deep-water forms. Of the Littoral zone, *i. e.*, the zone between high and low water, several facies may be recognized. The following are well marked on this coast: (1), the sand-beach facies; (2), the mud-flat facies; (3), the stony-beach facies; (4), the rocky-cliff facies; and (5), the bridge-pile facies. Each has its characteristic fauna, and minor subdivisions may be recognized in each. The Laminarian zone, 0 to 15 fathoms, can be explored only in its tide-pool facies or at extreme low water. The beaches, however, are often strewn with animals from this zone, brought thither by the waves during storms. Account will be taken of these zones in the descriptions which follow.

A. Sand-beach, stony-beach and mud-flat facies of the Littoral zone, with cast-up representatives of the Laminarian zone.

REVERE BEACH.

Route.—By rail, Boston, Revere Beach and Lynn railroad (narrow gauge); Atlantic avenue station, to Bath House station; fare, $.10.

By electrics, from Scollay square by Lynn and Boston railroad, Revere Beach car.

By bicycle, from Scollay square to Chelsea ferry, foot of Hancock street, *via* ferry to Chelsea, thence out Broadway to Beach street, which latter leads to the beach. Walk along beach and return from Point of Pines station, near the northern end of the beach.

Along the upper beach line, reached only by the storm waves, may be found usually an association of deep-water animals brought there during northeasterly storms. Among the molluscs, *Lunatia heros* may be mentioned as the commonest gastropod, the large white subglobose shell of this species usually being a conspicuous object. This shell by no means is confined to the upper part of the beach, but occurs all over it, and frequently is very abundant. Not uncommonly the living animal will be found in the shell, having been left on the strand by the retreating tide. The "egg collars" of this species are plentiful on the beach in mid-summer. Each consists of "a mass of sand glued together into the shape of a broad bowl, open at the bottom and broken at one side. Its thickness is about that of an orange-peel, easily bent without breaking when damp, and when held up to the light will be found to be filled with cells arranged in quincunx order. Each of these cells contains a gelatinous egg, having a yellow nucleus, which is the embryo shell."[1]

The larger empty shells of *Lunatia heros* are often repositories of a variety of more or less sedentary animals. *Hydractinia polyclina* not infrequently encrusts them, although the majority of specimens of the hydroids thus obtained are worthless. *Crepidula plana* and *C. fornicata* are to be found usually on such shells, the former on the inside. Young specimens often may be obtained. Our common limpet, *Acmœa testudinalis*, is found also clinging to these large gastropod shells. In addition, various species of encrusting bryozoans may be found on these shells. Shells of *L. heros* of all sizes may be obtained readily during a short walk on the beach, at almost any time. The small checkered variety *L. triseriata* is not uncommon on the lower portion of the strand.

A species which may be mistaken for *L. heros,* and which is found occasionally on this beach, is *Neverita duplicata.* This is distinguished by its more depressed spire, and by the strong callus which partially or entirely covers the umbilicus.

[1] Binney and Gould, pp. 339, 340.

Of other large gastropods, *Buccinum undatum* and *Chrysodomus decemcostatus* should be mentioned. These may be found commonly in the "storm zone," among the debris of the upper beach. Perfect specimens seldom are thrown up. Almost always they are much dissolved, and riddled by the boring sponge *Cliona sulphurea*. Calcareous algæ frequently cover them with an encrusting deposit. Hydractinia usually is found on those shells which have served as the home of the hermit crab. Perfect specimens are common on the coast of Maine, but seldom are seen on the Massachusetts beaches. The egg cases are found occasionally on our shores. With the two preceding species may be noted at times the smooth fusiform shell of *Neptunea curta* (*Fusus islandicus*), but this is rare on Revere beach. *Purpura lapillus* is found on the beach, but belongs properly to the rocky shores.

Two species of Crepidula are common on the strand. These are *C. fornicata* and *C. plana*, the latter usually attached to other gastropod shells. *C. fornicata* is the more common, and specimens of considerable size and of all degrees of convexity may be obtained. The coloration varies to some extent, as do also the thickness and curvature. Frequently a series of shells may be found attached to one another, making a solid pile; but this is more frequently met with on mud flats. The young of this species and of *C. plana* show a beautifully coiled embryonic shell or protoconch.

Among the smaller gastropods always found abundantly on the lower part of the beach, the commonest is *Tritia trivittata*. This shell is recognized easily by its turreted spire and the strong vertical varices cancellated by revolving lines which ornament its surface. It occurs by the thousand, sometimes heaped up into ridges parallel to the wave front, sometimes spread out over large areas of the beach. Many of the shells will be found with a round hole bored through them; and this condition has been attributed to *Lunatia heros*, one of the most voracious gastropods of our shores (Gould).

A somewhat larger, less delicate, and less strongly ornamented shell, *Ilyanassa obsoleta*, may be found commonly associated with *Tritia trivittata*, although it is more at home on the muddy shores of estuaries and bays. With these occurs the small smooth *Lacuna vincta*, usually in great numbers. It is seldom found alive on the beaches, but living animals may be taken from the roots of the Laminaria and other seaweeds, as well as from stones dragged up by storm waves. The shell is recognizable by the general smoothness of its surface, and by the peculiar crescentic groove which leads downward from the umbilicus, parallel to the inner lip.

One of the best methods of collecting these small species, as well as the small pelecypods associated with them, is to skim off the surface layer of sand and shells, to be sorted at leisure. In this way much valuable time will be saved.

Littorina littorea is common on the more stony portion of the beach south of the Bath house, but is found frequently on the sandy part of the beach. This species, as is well known, is not a native; but has been introduced from the west European coast, coming to Massachusetts probably by way of Halifax, N. S.

Of pulmonate gastropods, the little *Melampus bidentatus* is the only one commonly associated with the molluscs on the beach. The shells of this species are more abundant near high-water mark than on the strand. The animal always may be found in the marsh lands behind the beach, where the tide occasionally overflows. In the marshy lands about Oak Island, near the centre of the beach, it is a common form. It may be recognized by its resemblance to Oliva, by the two folds on the inner enameled lip, and by its thin translucent character and brownish horn color.

Of the polyplacophora, *Chiton* (*Trachydermon*) *ruber* is the only one likely to be found. This must be looked for among the roots of the cast-up Laminaria.

Pelecypods occur more numerously and in greater variety. Along the upper part of the beach *Cyprina islandica* almost always may be found, brought up from deep water during storms. When fresh the shells are covered by chestnut-brown epidermis, but after a period of exposure this is worn off. In addition to the epidermis, the subcircular outline of the valves, strongly forward-pointing beaks, dark, nearly black, external ligament, and absence of pallial sinus are distinctive marks. *Mactra solidissima* is another large pelecypod, common along the upper part of the beach. The animal lives in the sand below low water and can be dug at very low tide. This is the giant, beach, or dipper clam of the fishermen, and is esteemed as an article of food. Its large size, transversely ovate, somewhat triangular form, and peculiar spoon-shaped ligamental area separate it from all shells except *M. ovalis*, which is frequently, though more rarely, found in similar situations. This species may be distinguished by its smaller size, coarser surface, thick coarse epidermis of a dusky brown color, and less convexity of valves.

Another large and handsome shell, found occasionally on Revere beach, is *Thracia conradi*, distinguished by its peculiarly constricted posterior end, toothless hinge, and disparity in size and convexity of the two valves. It is found only after violent northeast storms, but probably lives buried in the sand a little below low water (Gould). *Zirphæa crispata* also is found occasionally after storms. This is one of the most striking shells of our shore, distinguished by the highly ornamented anterior and smooth posterior portion of each valve.

The most abundant shell on this beach after a northeasterly storm is the razor shell, *Ensatella americana*. The animal lives in the sand and mud above and below low water, and may be obtained alive by digging at low tide. The shells may be found by the thousand in spring, or after autumn storms, on the beach near Oak island station, where they are present in all sizes. Associated with this species usually occurs *Siliqua costata*, although this is commonly less abundant. It looks somewhat like a short Ensatella, and is recognized by the color of its epidermis, which is yellowish green blended with livid violaceous (Gould); and by the two whitish rays, one passing backward, the other nearly straight downward. The strong rib on the interior of the valve is also a distinctive mark.

One of the most beautiful species found on this beach is the little *Solemya velum*, the epidermis of which projects beyond the edge of the shell like a curtain, with a scalloped margin. It is especially abundant early in the year, and is to be sought among the seaweed and shells near the water's edge at low tide. The larger and less convex *S. borealis* is occasionally also found on this beach and is distinguished by its size and grayish blue or lead color.

Another beautiful shell, occasionally abundant, is *Lyonsia hyalina*, readily recognized by its delicate pearly and translucent character and radiating surface striæ. It is to be sought at low water in the sand and among seaweeds. Binney and Gould state that "in April, 1836, the beach at Chelsea [Revere beach], was strewn with multitudes of very large and mature" shells.

The shells of the common mussel, *Mytilus edulis*, and the horse-mussel, *Modiola modiolus*, are common upon this beach; the former carried up from mussel-beds off shore, the latter commonly brought from deeper water, held firmly in the roots of the Laminaria. The latter species is often very large and robust, and frequently shows deformations caused by obstructions during growth. The plicated *Modiola plicatula* is much less common on the sand beaches, but may be found always among the marsh grass near the mouth of fresh-water streams.

Several other species of molluscs occur among the roots of the Laminaria, in addition to a number of other deeper-water animals. The shells most frequently found are *Anomia aculeata* and *Saxicava arctica*. Among the larger shells occasionally thrown on to our beaches are *Petricola pholadiformis*, readily recognized by its plicated anterior end; *Astarte castanea*, known by its smooth chestnut-brown epidermis; and *Cytherea convexa*, a shell very similar to *Venus mercenaria* of the south coast, but differing in its chalky white color, smooth surface, and more rotund form. *Venus mercenaria* has been found, but rarely. *Macoma fusca*, a thin, white, transversely subovate shell, with a dusky epidermis, occurs on this and other sand beaches. Its habitat is shallow still water, with a muddy bottom.

Among the small shells found in great abundance on our beaches, *Tellina tenera* and *Tottenia gemma* are exceedingly common. The former, a tenuous semi-translucent shell with short pointed posterior end, often more or less colored, and of a pearly lustre, is usually the most abundant. Single valves and shells with the two valves united are found almost everywhere on the surface of the sand. Many of these show a hole bored by some carnivorous gastropod. The animal is found sometimes on the beach, but its habitat is probably in the fine sand just below low water. *Tottenia gemma* may be recognized by the amethystine color of its umbonal region. It is a very pretty little shell, and early attracted attention, according to Gould having been sent to England as one of the natural curiosities of this coast. Its small size causes it to be overlooked, unless the method of collecting suggested for the small gastropods be employed. It is thought commonly to be the fry of *Venus mercenaria* on account of its purple beak (Gould). With these species are found many young specimens of Mya, Mactra, Mytilus, Modiola, etc. *Mya arenaria*, *Mya truncata* and *Pecten tenuicostatus* also occur; the first in abundance, the second rare.

Next to molluscs, crustaceans and worms are the most abundant forms of life on this beach. Under the decaying seaweed of the storm-wave zone, as well as farther down the beach, are always immense numbers of the leaping amphipod *Orchestia agilis*, commonly called "beach-flea" or "sand flea." In color this species is dark olive-green or brown, resembling the decaying seaweed under which it lives. It is a very active leaper, the leaping being effected by means of the posterior appendages. It burrows in the sand under the seaweed, like its light colored and blue-eyed ally Talorchestia, which inhabits the lower part of the beach in vast numbers on outer Cape Cod and southern shores, where its burrows usually are very abundant near the water's edge. A much larger, quite closely related species, *Gammarus ornatus*, is found low down on the beach under stones, especially where fine mud has taken the place of the sand. This amphipod does not leap; but when disturbed, actively wriggles away on its side, by bringing head and tail near together and then straightening out suddenly, while bracing itself by the posterior abdominal appendages; thus repeating in a milder manner the leaping motions of its more active relatives. I have seen large individuals grasp the smaller ones by the middle of the back with their posterior thoracic legs, and quickly scud away to hide under stones or in the mud. In water these animals swim usually on their sides or backs. This species is olive or dark brown in color. On sandy flats and in pools left at low tide the common sand-shrimp, *Crangon vulgaris*, is always abundant, either swimming, resting on or buried in the sand, tail end down. Another abundant small crustacean on this and other sandy shores is the isopod *Idotea cœca*. The animal lives generally just below the surface of the sand, in which position it moves about, leaving a meshwork of curious "trails" all over the beach. This is seen well on the lower flats of fine sand, which are covered almost entirely with such markings. They can be seen even better on the first beach at Swampscott. The animal is to be found usually at the end of a trail, its whereabouts being marked by a little lump of sand. Two species of " hermit crabs," *Eupagurus bernhardus* and *E. longicarpus*, may be found dead occasionally on the beach after storms. The former inhabits the shells of Lunatia and Buccinum, while the latter, a smaller species, usually lives in shells of *Tritia trivittata* or other small gastropods. The former often has its shells covered with Hydractinia, although, as already noted, most of these will be found dead. The most common crab is the cosmopolitan *Cancer irroratus*, which is found both dead and in a living state. It is most abundant among the rocks and in the rocky pools. The horse-shoe crab, *Limulus polyphemus*, is not common on the Revere shore; although the exoskeletons and dead animals, as well as living ones, are cast up occasionally on the beach, or stranded by the retreating tide. It is seldom that good specimens can be obtained, these being far more common on the flats and marshy lands in the southern part of Massachusetts bay, and still more so on the South and Cape Cod shores. Occasionally the animals may be seen swimming below low-water mark on the Revere shore. Among the cirrhipeds several species of goose barnacle, Lepas, and the acorn barnacle, *Balanus balanoides*, are common, the latter encrusting stones on the stonier portion of the beach.

Of nemertean worms, *Meckelia ingens*[1] may be obtained opposite Oak island station, in muddy sand one to two feet below low-water mark. In the same place the polychæte worm *Nereis virens* occurs, at a similar depth below low water. With it live several species of Rhynchobolus. In the muddy sand on Saugus river at the northern end of the beach, and a little beyond the Point of Pines station, several polychætes may be found by digging. Among these are *Climanella torquata*, which constructs long round tubes of agglutinated sand, and *Rhynchobolus americanus*. With these worms occurs the holothurian *Leptosynapta girardii*, usually in considerable numbers. The sedentary worms are represented by several species of Spirorbis, which are found in abundance attached to seaweeds cast up on the beach by storms. The animals frequently may be obtained alive just after they have been cast up.

Echinoderms are not very common on the beach. Small starfish, Asterias, may be found commonly among the roots of Laminaria where small sea urchins, Strongylocentrotus, and brittle stars, Ophiopholis, also occur. The sea urchins are found also in a more or less battered condition among the seaweed after storms. The sand dollar *Echinarachnius parma* rarely is found on this beach. Two species of holothurians also may be found on the beach after storms. Of these, *Caudina arenata* often occurs in great

---

[1] For notes upon this and the following worms, I am indebted to the Zoölogical Department of Harvard College.

numbers, and may be dug also at low water a little below the surface of the sand. *Leptosynapta girardii* may be dug on the banks of the Saugus river at the northern end of the beach, as well as in front of the Point of Pines hotel. Balanoglossus may be obtained by digging near low-water mark on the beach. It is most abundant opposite Oak island, and from there to the Point of Pines. After northeast storms it frequently is cast up in numbers on the beach.

The two sponges common on this beach are *Cliona sulphurea* and *Chalina oculata*. The former is represented chiefly by its borings

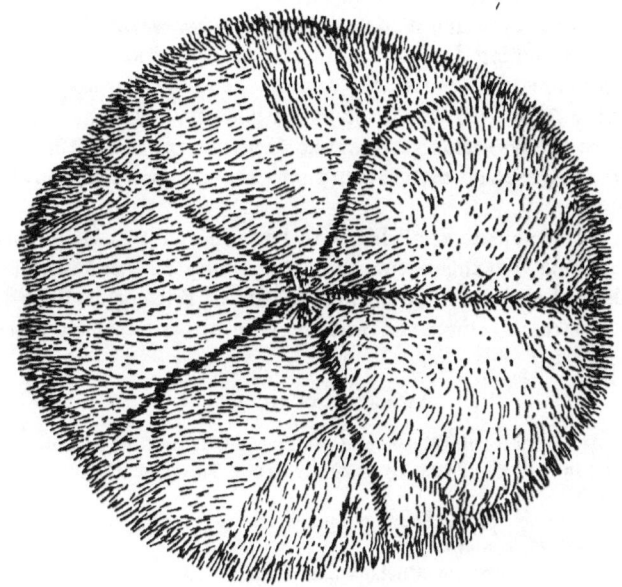

Echinarachnius parma.
(After Fewkes, by courtesy of the Essex Institute.)

in the shells, and the latter by dried specimens on the upper part of the beach.

A number of hydroids and bryozoa are cast up attached to various seaweeds. Among the former *Obelia geniculata* may be found commonly attached to the fronds of the Laminaria, although specimens in the living condition seldom can be obtained.

Sertularia pumila usually is found attached to the rockweed Fucus and Ascophyllum, and *Sertularia argentea* can be found also among the seaweeds of the beach, between Oak island and the Point of Pines. This species is especially abundant after an easterly storm, and specimens in good condition may be obtained. It is readily recognized by its long slender form and alternating hydrothecæ. *Hydractinia polyclina* frequently covers the dead shells inhabited by hermit-crabs; but the specimens obtained on the beach are either dried, showing nothing but the polyparium, or else the animals are white, an indication that they are dead. Very seldom can living specimens be obtained on the beach. *Clava leptostyla* may be found attached to Fucus on the rocky portion of the beach, near Crescent Beach station.

Of bryozoa, the genus Membranipora is represented by several species, encrusting Fucus, Laminaria, or *Chondrus crispus*. The most common species is *M. pilosa*. *Crisia eburnea*, a small branching form with slender stellate branches, and a coarser form, *Cellepora ramulosa*, also are common on the seaweed. Other bryozoa are to be seen on the cast-up seaweed, and sometimes on stones and shells.

The only tunicate at all likely to be found on this beach is *Boltenia rubrum*, but this occurs more frequently on Swampscott and Marblehead shores. The normal habitat of this species is in from two to fourteen fathoms of water, where it is attached to rocks or shells.

SWAMPSCOTT BEACH.

Route. — *By train*, Boston and Maine road, from North Union station to Lynn; fare $.20. From Lynn *via* Lynn and Swampscott electrics to the first beach.

By electrics, from Scollay square, in the Lynn and Swampscott cars; fare $.25, round trip.

By bicycle, through Chelsea and Lynn to the beach, following car tracks, 15 miles.

A large number of the species found on Revere beach may be obtained also on Swampscott beach. In addition, a number of others, especially shells, may be found, which are largely derived from the stomachs of fishes cast on the beach by fishermen. One of the most beautiful of these is the small gastropod *Margarita obscura*. This, although a small shell, is conspicuous from its pearly lustre and iridescence, which appears when the exterior is worn off. Other species of Margarita also occur. *Neptunea curta* is found more often here than on Revere beach, and perfect specimens can be obtained frequently. *Petricola pholadiformis* is present on this beach in considerable numbers. The isopod *Idotea cœca* is very common, as are also other species of Idotea. The strand sometimes is covered with the curious trails made by the first mentioned species.

Many of the shells are more perfect than on Revere beach, but the variety is not so great.

NAHANT NECK.

Route. — *By rail*, Boston and Maine road from North Union station to Lynn; fare $.20. By barge or electrics to beach.

By electrics, from Scollay square to Lynn (Lynn & Boston cars); change to Beach car.

By bicycle, through Chelsea to Lynn, 12 miles; follow car tracks to beach, 1 mile.

This bar, reaching southward from Lynn and tying the Nahant islands to the mainland, has an open sea-beach and a harbor-beach, giving a difference of fauna on the opposite sides. In spite of this fact, the animal life found here differs comparatively little from that of Revere beach; for the great length of that beach and the diversity in character of the material composing it produce a great variety of physical environments, and hence cause a diversified fauna.

Among the rarer shells found on Nahant beach after storms are *Glycymeris siliqua*, *Ceronia arctata* and *Pecten tenuicostatus*. The first of these is recognized readily by its elongated oval form and its black, dense and shining epidermis. *Ceronia arctata* may be known by its short posterior end, and *Pecten tenuicostatus* by its large size and smooth surface.

Worms and other burrowing animals are plentiful on the Lynn bay side of the neck. Among gephyreans, *Phascolosoma gouldii* always may be found at low tide by digging in the sandy mud. Of polychætes, *Clymenella torquata* is found also in the mud of this beach at low tide. *Amphitrite ornata* is one of the most beautiful marine worms living on these beaches. It may be obtained in the mud at low tide, the position of the animals being marked by the holes in the sandy mud.[1] It may be recognized by its blood-red gills and numerous long flesh-colored tentacles. It constructs tubes from mud and sand. Several species of Nereis occur also

---

[1] Noted by Dr. Parker and Dr. Davenport.

in the mud of this beach. *N. palagica* has been obtained from the sand at about high-tide mark, and *N. virens* may be found also, lower down on the beach.

B. *Estuarine faunas. Muddy beach and bottom, bridge-piles, mud-flats and rocky shore.*

BEVERLY.

Route.— *By train,* Boston and Maine road from North Union station to Salem; fare $.35. Take electrics (or bicycle, two miles) at Essex street for Beverly bridge; stop at Salem end of bridge. Obtain boat from Mr. Powers, No. 2 Bridge street. Chief collecting grounds under the bridge, and on the ledges in the estuary. The time of lowest tides should be selected.

In the mud about the boat-landing, *Ilyanassa obsoleta* is the commonest gastropod. It may be seen crawling by thousands in the mud, and can be recognized easily by the comparatively smooth spire and rather coarse aspect. With it occur large numbers of *Littorina littorea.* Hermit crabs are very abundant in the mud about the landing, but in this part of the estuary they seldom bear Hydractinia. *Gammarus ornatus* may be found also, although perhaps less commonly than on the more stony portions of our shores. The best collecting is from the piles of the bridges, especially the road bridge. A number of distinct zones may be recognized on the piles, each with its characteristic type of animal life. The highest zone is occupied by the common barnacle, *Balanus balanoides,* which completely covers that portion of piles, and in a measure serves as a protection to them. These animals grow so closely crowded together that they are unable to assume their normal form, and are obliged to increase chiefly in length. One to two inches in length is not uncommon in this location, with the diameter of the corona seldom exceeding half an inch. Next below the barnacle zone we find the timbers, where shaded, covered by our commonest campanularian hydroid, *Campanularia flexuosa.* This is flesh-colored, and often occurs in such luxuriance as to cover the piles completely. When the tide is out these hydroids hang down in brownish masses, often closely adhering to the wood; and under such circumstances they offer no indication of the beauty which they will exhibit as soon as the returning tide revives them. The species is quite hardy, and may be kept alive easily for several days, provided it has a sufficient amount of water. It seems to require periodical exposure, for we seldom find it growing much below low-water mark. The same species inhabits the piles of all our bridges across tide-streams and estuaries, often occurring where the water is very muddy. It lives also on exposed shores, growing on and under the rockweed and in the fissures of the rock, wherever protected from the force of the waves.

A little lower down the piles are covered thickly with mussels, *Mytilus edulis*, which in turn frequently are covered by the *Campanularia flexuosa*. This grows on the mussels in thick bunches, from which it is more easily obtained than from the timbers.

Almost at low-water mark, and exposed only at the lowest tides, we find another campanularian, *Obelia commissuralis*, which is occasionally abundant, although one usually must search for it. It is distinguished easily from *Campanularia flexuosa* by the longer, more slender, and more profusely branched form of the hydrozoarium. *Obelia gelatinosa*, a somewhat coarser species but similar in its general form, also may be found in this position. In the same zone we find the beautiful tubularian, *Parypha crocea*, which is often very abundant. This is the largest and perhaps most beautiful of our common hydroids; and is easily recognized by the large heads, the two circles of tentacles, and the brilliant pink color. Not infrequently the heads are broken off, the stems remaining like bunches of stubble. Nevertheless they retain their vitality, and probably have in company with other tubularians the power of regeneration. This species grows very rapidly, and will become attached to objects submerged only a short time. A dory which had been for two weeks in the tide-way near the draw of Beverly bridge in June, 1897, had its bottom covered completely with this beautiful hydroid, most of the individuals apparently in an adult stage.

Below the zone of tubularians, the piles are covered with sea-anemones, *Metridium marginatum*. These grow so thickly that in places there is hardly a space between them. They vary greatly in color, but the majority are reddish-brown. At the lowest tides many of them are exposed, and then hang down limp and unsightly. Among them, as well as among the hydroids, nudibranchs are not uncommon. *Dendronotus arborescens*, with two rows of large arborescently-branched gills on its back, is the most abundant. This species feeds on hydroids, and its narrow foot adapts it eminently for creeping over these small creatures. It will be found commonly in the collecting bottles, after they have stood for a time; but is not noticed generally amongst the hydroids on the piles.

On the fronds and among the roots of Laminaria, which are frequently caught on the piles, many interesting species may be

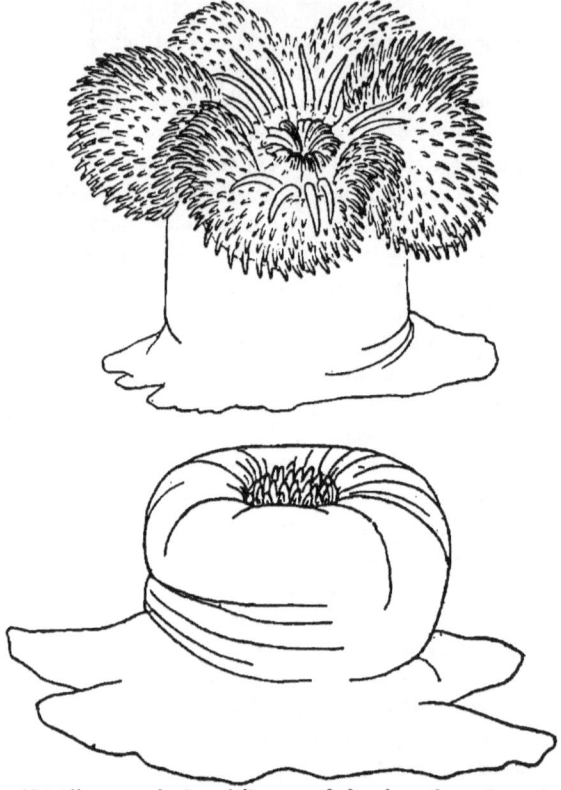

Metridium marginatum fully expanded and partly contracted.
(After Fewkes; through the courtesy of the Essex Institute.)

obtained. *Obelia geniculata* often covers the fronds thickly, giving their surfaces a downy appearance when floating in the water. This species is a native of the cold water of the outer shore and is seldom obtained alive except in its natural habitat. The drainage-polluted water of the Beverly estuary is especially unfavorable to its existence, although it seems to be the natural habitat of a number of closely related species. On the stems of Laminaria, various bryozoans as well as Spirorbis commonly are found; and *Parypha crocea* sometimes covers them thickly. Among the roots, besides many deep-water shells, the more delicate species of hydroids are not uncommon. *Clytia poterium* and *C. bicophora* are frequent. Their usual habitat is in rocky tide-pools where the water is clear and cool. They are attached almost invariably to seaweeds, or the stems of other hydroids; and, according to Agassiz, never to solid rock or to immovable substances. *C. poterium* has the summits of its campanulate calices smooth, and the gonophores

Parypha crocea.   *Corymorpha pendula.*   *Clava leptostyla.*
(After Fewkes, by courtesy of the Essex Institute.)

arise from the creeping stolon. The pedicels of the nutritive zoöids are strongly ringed from top to base. *C. bicophora* is distinguished from the last by the notched margins of the calices, the annulated gonophores which arise from the stolons or the stems, and the weakly ringed condition of the pedicels of the nutritive zoöids.

On the ledges near the mouth of the estuary several species of hydroids and other littoral animals always may be obtained. Among the former is *Clava leptostyla*, which can be found here in abundance at any season of the year. It is attached to Fucus and Ascophyllum, and more rarely to the rocky bottoms of the pools. In the former case it is exposed at low tide, commonly for several hours at a time. It is usually of a deep flesh color, which in exposed and contracted animals appears much deeper than in expanded ones. *Sertularia pumila* is another hydroid abundant under and on the rockweed; and *Campanularia flexuosa*, though much rarer in this habitat than in the preceding, also is found. Barnacles are common, and near them occurs *Purpura lapillus* in considerable numbers and in great variety of form and coloration. The pear-shaped egg-cases of this species may be found usually in clusters attached to the rock or seaweed. They are white or yellow, turning with age to pink or purple. Three species of Littorina always can be found on these rocks. The largest, *L. littorea*, is extremely abundant and, like all littoral species, shows considerable variation in form and color. The small high-spired species, *L. rudis*, also is common, and likewise shows much variation. The spotted variety, *L. tenebrosa*, also occurs. *L. palliata*, another common variety, has a low spire and is more globose and smooth. It is sometimes yellow or orange in color; but more commonly dark brown or olive, not infrequently striped, banded or spotted.

In the shelter of the ledges is an extensive mussel-bed, and starfish are not uncommon. They are more abundant however in the immediate vicinity of the ledges, where the large *Asterias vulgaris* may be seen by the hundred on the muddy bottom and among the eel-grass. *A. forbesii* also occurs here, but is more common on the ledges.

The smooth starfish, *Cribrella sanguinolenta*, sometimes is found on the ledges; but on the whole it is a rare species on our coast. Clinging to the ledges and to submerged stones and shells, the common limpet, *Acmæa testudinalis*, can be found in abundance. It varies somewhat in shape and coloration. On the muddy bottom accessible at the lowest tides, where the water is only from one to two feet in depth, the large hermit crab *Eupagurus bernhardus* is very abundant. It inhabits the shells of *Lunatia heros*, *Buccinum undatum* and large shells of *Littorina littorea*. Upon these shells may be found commonly the polymorphous hydroid *Hydractinia polyclina*.

On the eel-grass the curious "no-body-crab," Caprella, may be found; various species of hydroids as well as gastropods, similar to those found on the ledges, occur clinging to the eel-grass.

In the deeper water of the estuary, the large simple hydroid *Cory-*

morpha pendula has been found. The brachiopod *Terebratulina septentrionalis* also has been taken, farther out.

During the summer months, the common white jelly fish, *Aurelia flavidula*, always may be seen in the water near the tide-way. The larger brown *Cyanea arctica* also is seen occasionally.

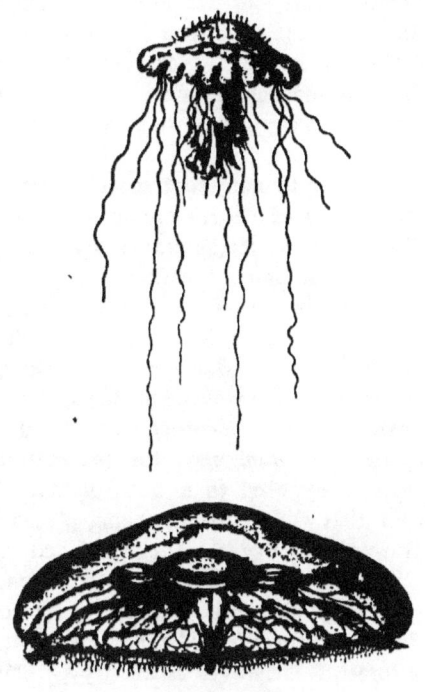

Young Cyanea arctica.
Aurelia flavidula.
(After Fewkes, by courtesy of the Essex Institute.)

Rocky shore faunas, comprising the rocky cliff facies of the Littoral zone, and the tide-pool facies of the Laminarian zone.

The rocky cliff facies of the Littoral zone is characterized everywhere by the abundant growth of the rockweed Fucus and Ascophyllum. (See page 97.) On and under these rockweeds many animals find a congenial place of abode, in situations that would appear all but conducive to their existence. It is true that the rockweed breaks the force of the wave-impact, which otherwise would be absolutely destructive to all but the most protected of the littoral animals; nevertheless one cannot cease to wonder that animals apparently so frail as the hydroids and bryozoans can withstand the continued beating of the surf, even with the protection afforded by the rockweed. This applies especially to such unprotected hydroids as *Clava leptostyla*, which may not infrequently be found on the rockweed in the most exposed places. The reason for the luxuriant development of animal life between tide marks on exposed shores is no doubt, as often has been pointed out, the large supply of food furnished in a ground-up condition by the waves.

It will be interesting to note that most of the common littoral animals of these exposed shores are in some way or other protected from the full force of the waves by their thick shells or other covering. Those same coverings serve them also as protection against the drying effect of sun and air during their periodic exposure between tides. Among hydroids, *Sertularia pumila*, always found in abundance on the rockweed, is best protected from the force of the waves and the heat of the sun by its operculated hydrothecæ. In consequence of the closure of the hydrotheca by the operculum upon the contraction of the polyp, the hydroid can be exposed to a hot July sun for five or six hours, and after that will revive on being placed in water; while the same exposure will dry the rockweed and render it brittle. The other common hydroid of the exposed shores, *Campanularia flexuosa*, has its polyps protected likewise by hydrothecæ; but not having the additional protection of the opercula, it is less able to withstand the force of the waves or the drying effect of exposure to the sunlight. In consequence, the usual position of this hydroid is under the rockweed, upon its stems, or on the ledges. Thus it is not only protected from the force of the waves but kept moist by the rockweed, which at low tide covers it like a moist curtain. The common bryozoan of these exposed shores is *Alcyonidium hispidum;* the soft, thick and furry polyparium of which is sufficient protection against the force of the waves and the drying power of the sunlight, especially as this species grows chiefly around the stems of the rockweed, where it is in no danger of being beaten against the rock.

The gastropods of the exposed rocky shores are protected by thick shells. This is shown by the commonest species inhabiting this zone, *Purpura lapillus*. The shells of Littorina, though not comparatively as thick as those of Purpura, nevertheless are strong and capable of resisting a great force. The barnacles are protected admirably by their corona and the well-fitting valves of the inner shell. The mussel, *Mytilus edulis*, while perhaps sufficiently protected by its shell, seems nevertheless to be incapable of sufficiently firm fixation to withstand the force of the waves. Consequently we find it only in sheltered nooks behind rocks, or under the rockweed ; in all of which positions its habit of growing in closely packed beds helps to secure stability of attachment for the individual.

The following two localities, both near together and within easy reach from Boston, are selected to illustrate the littoral fauna of the cliffs, because they furnish also beautiful examples of tide-pools. Other headlands, such as those of Marblehead Neck and Cape Ann on the north, and the Nantasket cliffs on the south, furnish good illustrations of the littoral fauna of the exposed ledges ; but they are not so rich in tide-pools. Nevertheless these ledges deserve close attention; as the degree of exposure, the character of the cliffs, and the force of the waves vary sufficiently to produce at least some variation in the fauna.

CASTLE ROCKS, NAHANT.

Route. — From Lynn (see pp. 11-13), barge or bicycle across Nahant neck to Castle rocks, about 5 miles. Barge fare, $.25 round trip.

A large number of interesting invertebrates always may be found in the tide-pools of Castle rocks, as well as on and under the rockweed which is plentifully exposed at low tide. In this latter habitat the two common hydroids of our rocky coast, *Sertularia pumila* and *Campanularia flexuosa*, are always abundant, not infrequently growing on the rocks under the seaweed, or in the crevices. The egg-cases of *Purpura lapillus* and other gastropods are also common under the rockweed, while the animals themselves may be found everywhere, especially where the barnacles upon which they feed are abundant. The shells of the Purpura vary considerably in size, form, thickness, ornamentation and coloration ; and in a comparatively short time a large number of specimens may be gathered ranging in color on the one hand from pure white to pure orange, purple or dark gray, and on the other from variegated and mottled to regularly striped and banded. Nearly smooth varieties occur side by side with those coarsely ribbed, and occasionally an individual with strong varices may be found. Variations in elevation of spire and outline of peristome also occur. The two small species of Littorina, *L. rudis* and *L. palliata*, are always abundant on the rocks or seaweed, both showing considerable variation in form and color. Both species are vegetarians living upon the seaweed (Verrill). The first of these is of special interest from being viviparous (Verrill). Around the stems of the rockweed, on the more exposed ledges, where the cool clear water bathes them the [greater part of the time, our common furry bryozoan *Alcyonidium hispidum* usually may be found. When the polyps are withdrawn the bryozoarium is not very attractive; but when the specimen is placed in a jar of clear cool water the pale purple polyps soon expand and the colony becomes an object of great beauty.

In the fissures left by worn-out dikes, as well as under the overhanging portions of the ledges, various sponges of the genus Halychondria (?) may be found. The most abundant of these have a light greenish tint, and occur also in the tide-pools. With them lives a bright red species.

Numerous small hollows or diminutive tide-pools may be found on these ledges, and they furnish the best opportunity for observing the barnacles in the operation of feeding. In order to see this, one must get down close to the pool, in such a position as not to obstruct the sunlight. Unless the water has become too warm, the little creatures almost always will be found active. In these small pools, various isopods and amphipods are common.

The larger tide-pools, always fringed with rockweed and lined with Corallina, *Chondrus crispus*, Ulva, and the long-streamers of the smaller laminarians, are the ideal collecting grounds for the student of marine invertebrates. A few interesting ones occur at Castle rocks, but there is probably no place on the Atlantic coast where tide-pools are so abundant and so rich in life as at East point, Nahant. The larger tide-pools of Castle rocks always are well stocked with *Modiola modiolus*, star-fish, and sponges; and in addition to these a number of the more delicate invertebrates occur. Among the rock masses, where protected from the force of the waves, the common sea-anemone, *Metridium marginatum*, lives in abundance. When the tide is out these animals can be observed readily in their native habitat where, in the shadow of the castellated cliffs, they seem not to be affected by the drawing off of most of the water, but remain with tentacles spread and bodies fully expanded, to the delight of the observer. When the water is warmed by the sun, the animals contract. This species is quite hardy, and can be kept in an aquarium for weeks, provided the water is kept cool and is changed every few days. It may be fed on bits of meat, snails, etc.

The frondose bryozoan *Bugula turrita* occurs abundantly in these tide-pools, and can be distinguished readily from the hydroids associated with it, by the flat character of its branches. This species shows the avicularia well; and under the microscope these may be seen active in a living specimen, if kept in salt water. *Membranipora pilosa* and other species of encrusting bryozoans are found on the branches of the rockweed, as well as the *Chondrus crispus;* and with these on the latter alga, may be found commonly the spiral shell of the sedentary worm *Spirorbis borealis*. Hydroids are abundant, but they are mainly campanularians. *Campanularia flexuosa* is not an uncommon species on the rocky walls of the tide-pools, but is more frequent under the rockweed about the rims of the pools. *Sertularia pumila*, most at home on the rockweed, nevertheless occurs in the pools, although much less commonly than on the rockweed. The two delicate species of campanulate hydroids, *Clytia bicophora* and *C. (Orthopyxis) poterium*, are also common in these pools; and with them occur numerous other animals, which are found likewise in the tide-pools of East point.

EAST POINT, NAHANT.

Route.— From Lynn by barge or bicycle, across Nahant neck to the Lodge estate. Follow foot-path to cliff.

At East point, Nahant, the cliffs are composed of metamorphosed slates and limestones with several intrusive beds (see pp. 10–13), the whole dipping inland at a high and practically uniform angle. Owing to the varying hardness of the beds, and the erosion of the softer in preference to the harder, a number of more or less completely enclosed tide-pools are formed at various levels, all of them containing an abundance and variety of marine organisms. The best of these are in the vicinity of Pulpit rock, and are accessible at low water. The margins of these crystal pools are lined with the drooping rockweed, and in their depths may be seen a luxuriant growth of the more delicate algæ, especially Ulva and *Chondrus crispus*, the latter reflecting beautiful blue and purplish colors from their oily surfaces wherever the sunlight strikes them. Coralline seaweeds are also very common, and among them the deep and clear-water mussel *Modiola modiolus*. Its surface commonly is covered with the pinkish encrusting Corallina, and

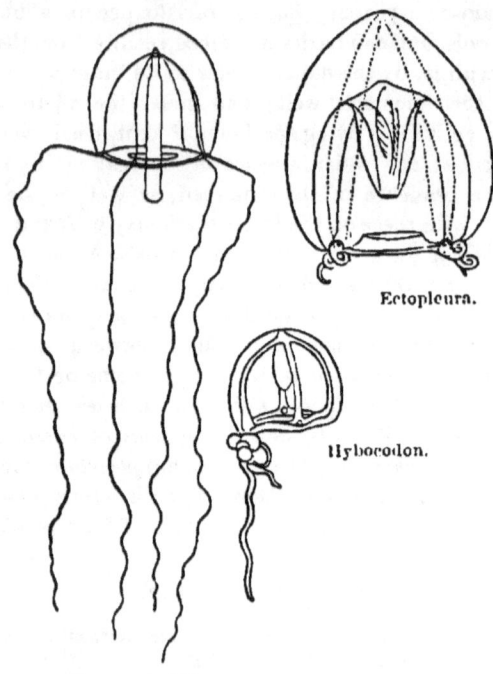

Coryne mirabilis.
(After Fewkes, by courtesy of the Essex Institute.)

rough with projecting epidermal spines; features which serve admirably to disguise the presence of the mollusc. Attached to these mussels we find not infrequently the more delicate species of hydroids, which are unable to withstand the exposure to which the more hardy littoral species are subjected. Among tubularians, *Eudendrium dispar* and *Coryne mirabilis* are most abundant. The medusæ of the latter species and of *Bougainvillia superciliaris* may be found commonly in considerable numbers in the water off the

Nahant headland as well as other parts of Massachusetts bay, during the spring and summer months. *Clava leptostyla* occurs on the rockweed in a number of places along the Nahant shore, and *Bougainvillia superciliaris* is a not uncommon inhabitant of the deeper tide-pools. Agassiz mentions the occurrence of *Hydractinia polyclina* on the rocks in these tide-pools—a rather unusual habitat for that species, which most commonly is to be found on shells inhabited by hermit crabs. The following tubularians are also mentioned by A. Agassiz[1] as having been found at Nahant: *Eudendrium tenue; Rhizogeton fusiformis; Syndictyon reticulatum; Gemmaria cladophora; Pennaria tiarella; Hybocodon prolifer*, and *Thamnocnidia tenella*. Among the campanularians, the genus Clytia is represented in these tide-pools by *C. intermedia, C. bicophora, C. (Platypyxis) cylindrica*, and *C. (Orthopyxis) poterium*. The last three of these species are common, the first has been recorded only by A. Agassiz. According to him, the following species also have been found: *Eucope diaphana* (= *Obelia diaphana*, Verrill); *E. attenuata* (= *O. geniculata*, Hincks; often on Laminaria in tide-pools); *E. polygona* (= *O. polygona*, Verrill); *E. parasitica; E. pyriformis; E. alternata; E. (?) fusiformis; Obelia commissuralis; Diphasia rosacea; Sertularia cupressina* and *Amphitrocha cincta*.

As in most places along our shore, so here the two hardy species, *Sertularia pumila* and *Campanularia flexuosa* (*Laomedea amphora* Ag.), are abundant on and under the rockweed which fringes the pools. Various medusæ often may be found in the pools; but the real fishing ground for these creatures is off East point, where they always may be found in number. " At this spot the sea actually swarms with life; one cannot dip the net into the water without bringing up Pleurobrachia, Bolina, Idyia, Melicertum, etc., while the larger Zygodactyla and Aurelia float about the boat in numbers."[2]

The attached medusa *Lucernaria (Haliclystus) auricula* is found sometimes on the seaweed of the pools; but its more normal habitat seems to be among the eel-grass, to which it may be found attached in many places along our shore.

Sea anemones are abundant in most of the pools, where they are shaded from the sunlight and sheltered from the strong waves.

[1] North American Acalephs.
[2] Agassiz, Elizabeth C. and Alexander. — Seaside studies in natural history, p. 86.

90 GUIDE TO LOCALITIES.

The sides of some of the pools not infrequently are found covered with small individuals.

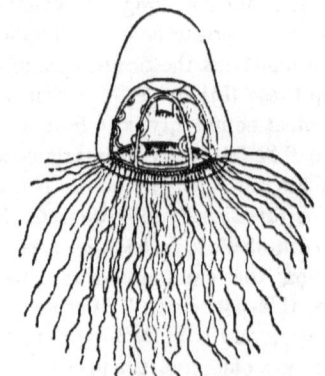

Melicertum campanula.
(After Fewkes, by courtesy of the Essex Institute.)

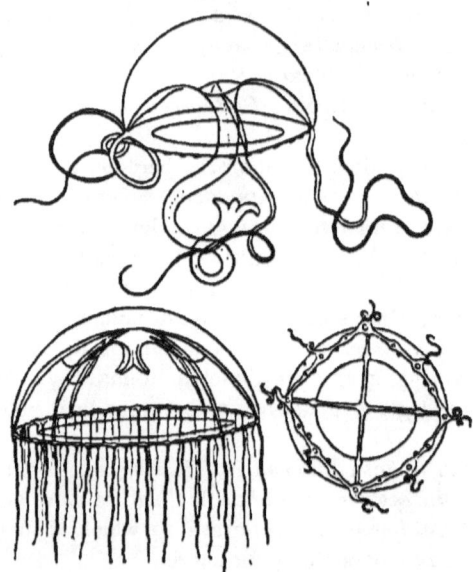

Staurophora.   Liriope.   Clytia.
(After Fewkes, by courtesy of the Essex Institute.)

Sponges are common in all the pools, the most abundant being a large smooth species of a delicate greenish hue. Bryozoa always may be obtained on this shore, the encrusting species *Membranipora pilosa* occurring on various seaweeds, while *Crysia eburnea* and *Cellepora ramulosa* occur most commonly on the Chondrus. *Bugula turrita*—often mistaken for a hydroid—is also a common species attached to the rocks or seaweed. *Membranipora pilosa* is recognized readily by the single layer of closely approximating oblique cells which are arranged in alternating order; and the rim of each is furnished with one long hair and several spinous denticles. *Crysia eburnea* is attached only basally and grows in the form of little white bushy tufts, much branched and often forming a round cluster. The cells are cylindrical in two rows, nearly opposite, and bend outwards with a gentle curve, terminating in a circular aperture. *Cellepora ramulosa* is a white calcareous dichotomously branched species, with irregularly clustered cells, each with a mucronate point on the margin of the aperture. It grows to a height of two or three inches (Johnston). *Alcyonidium hispidum*, always abundant on the open rocky shore, encrusts the stems of the Fucus; and with it occurs not infrequently the more delicate and smaller species *A. hirsutum*. These two are perhaps our most beautiful bryozoans when the polyps are fully expanded; although, when the polyps are contracted and the rough brown polypary alone is seen at low tide, it is one of the least attractive objects.

Spirorbis borealis, the common sedentary worm of this coast, always may be found in the tide-pools, attached to *Chondrus crispus* and other seaweeds. Many chætopods, nemerteans, and planarians occur also. The nudibranchs are represented by the common and beautiful *Dendronotus arborescens*, which is abundant in all the tide-pools. *Æolis papillosa* and *Æ. rufibranchiata* are found likewise among the seaweeds in these pools. Their egg-masses occur under small stones in the pools. Another gastropod abundant in these pools is our northern limpet, *Acmœa testudinalis*, which is found everywhere clinging to stones or shells. *Chiton* (*Trachidermon*) *ruber*, although more common in the deep water, from which it is brought up by the dredge, not infrequently occurs in these tide-pools. Clinging to rocks, and simulating in color the rose-tinted encrusting corallines so common on the stones and shells of these shores, it is distinguished with difficulty. Around the margins of the tide-pools and on the seaweeds and rocks exposed at low water, as well as crawling over the barnacle beds, is the most abundant coiled gastropod of these exposed shores, *Purpura lapillus*, of which the variations in color and form have been noted above. The varieties are nowhere so abundant as at Nahant, where brilliantly colored and variegated individuals are not uncommon. The three species of Littorina

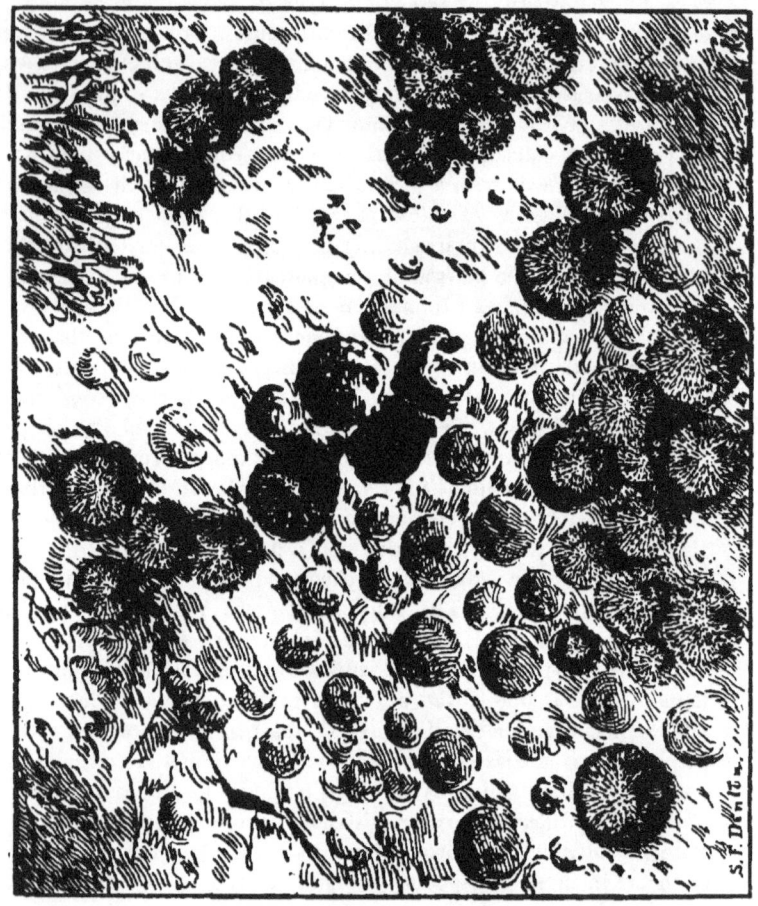

Sea urchins in excavations.
(After Fewkes, by courtesy of the Essex Institute.)

with their numerous varieties also are abundant. *L. palliata* and *L. rudis* are common on the seaweed and rocks near low water, while *L. littorea* may be met everywhere. The only pelecypods at all abundant on the rocky shores of this region are the mussels. *Mytilus edulis* is found in sheltered places, behind rocks and under seaweeds, growing between tides; while *Modiola modiolus* may always be found in the pools, where at first it is distinguished with

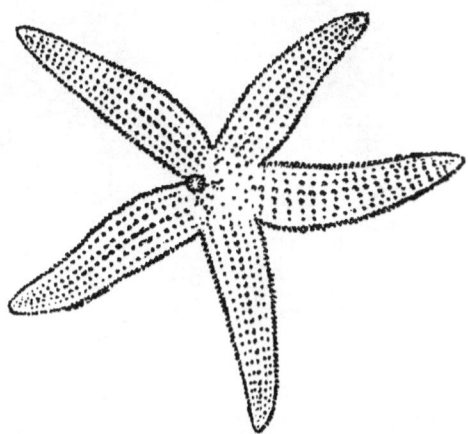

Asterias.
(After Fewkes, by courtesy of the Essex Institute.)

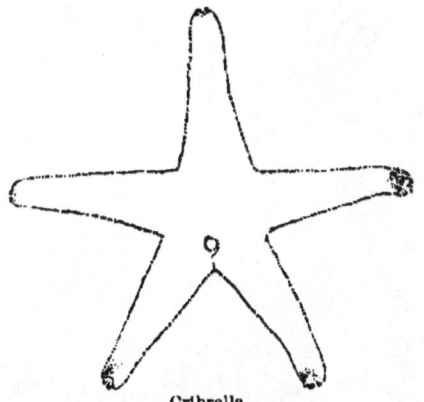

Cribrella.
(After Fewkes, by courtesy of the Essex Institute.)

difficulty because of the coralline growth upon it. Under stones in the tide-pools may be found also the small *Saxicava arctica*, a species much more abundant on the Maine coast.

The Echinoderms are well represented in the tide-pools. *Asterias*

Ophiopholis.

Strongylocentrotus dröbachiensis.
(After Fewkes, by courtesy of the Essex Institute.)

forbesii is abundant and varies greatly in color. *A. vulgaris* is also common. The smooth star-fish, *Cribrella sanguineolenta*, with only two rows of ambulacra in each ray, not infrequently is found, although on the whole it is a rare species on our coast.

The brittle star, *Ophiopholis aculeata*, is common in the pools, hiding away under stones and in crevices or clinging to the rocks under the seaweed. This species is much more common in the deeper waters off shore, where the dredge brings it up by the hundred. The common sea urchin of the north Atlantic coast, *Strongylocentrotus dröbachiensis*, is very abundant in the Nahant tide-pools. It may be found always in the deeper pools, hiding away in the corners and covering itself with shells, seaweeds, and other available protective coverings. This species is exceedingly abundant on some parts of the Maine coast, where it is uncovered by the thousand at low tide. It rests then among the stones, and is always more or less covered by dead shells, pebbles, etc.

Crustacea are not uncommon in the tide-pools, where several species of decapods always may be met. After these and the barnacles, the most prominent forms are the isopods *Idotea irrorata* and *I. phosphorea*. The latter is found not infrequently swimming in numbers in these pools. Both species appear to be more abundant farther north, on the coast of Maine, where they are among the principal inhabitants of the tide-pools on exposed shores. The two species are distinguished readily by the form of the pleon, which in *I. irrorata* has a tridentate termination, while in *I. phosphorea* it is pointed. Both forms are large, an inch or more in length, and both vary much in color, the chief tints being greenish and brownish. *I. irrorata* is commonly striped, while *I. phosphorea* is more often banded.

In addition to the animals mentioned from these tide-pools, a large number of others occur, including probably many undescribed species; thus furnishing an ideal spot for the student of marine invertebrates.

GENERAL REFERENCE WORKS.

1851. *Stimpson, William.* Shells of New England.
1853. *Stimpson, William.* Marine Invertebrata of Grand Manan. (Smithsonian contributions to knowledge.)
1862. *Verrill, A. E.* Revision of the Polypi of the eastern coast of the United States. (Bost. Soc. Nat. Hist., Mem., vol. 1, pp. 1–45, pl. 1.)

1862. *Agassiz, Louis.* Contributions to the natural history of the United States. Vols. 3 and 4.
1865. *Agassiz, Alexander.* North American Acalephæ. (Mus. Comp. Zoöl., Mem., vol. 1, No. 2.)
1865. *Agassiz, Elizabeth C.* and *Alexander.* Seaside studies in natural history. Marine animals of Massachusetts bay.
1866. *Verrill, A. E.* On the Polyps and Echinoderms of New England, with descriptions of new species. (Bost. Soc. Nat. Hist., Proc., vol. 10, pp. 333–357.)
1870. *Gould* and *Binney.* Invertebrates of Massachusetts (Mollusca).
1873. *Verrill, A. E.* Invertebrate animals of Vineyard sound. (U. S. Fish commission report, vol. 1.)
1880. *Hargar, Oscar.* Marine Isopoda of New England, etc. (U. S. Fish commission report for 1878, pp. 297–462, plates 1–13.)
1891. *Fewkes, J. Walter.* An aid to a collector of the Cœlenterata and Echinodermata of New England. (Essex Inst., Bull., vol. 23, Nos. 1, 2, 3.)
1892. *Verrill, A. E.* The marine nemerteans of New England and adjacent waters. (Conn. Acad. Sci., Trans., vol. 8, pp. 382–456, plates 33–39.)
1892. *Verrill, A. E.* Marine planarians of New England. (Conn. Acad. Sci., Trans., vol. 8, pp. 459–520, plates 40–44.)

III.

BOTANY: MARINE ALGÆ.

Prof. W. G. Farlow.

MARINE ALGÆ.

To those persons interested in botany who attend the sessions of the American Association, the coast of Massachusetts bay offers much that is attractive. Cape Cod forms the dividing line between the northern and the southern marine floras of our Atlantic coast. North of Cape Cod, if we except a few warm, sheltered coves, the species of marine algæ are those characteristic of arctic and sub-arctic seas and practically all the more striking forms found from Nahant northward extend to the coast of Greenland. The points most accessible and most likely to afford a good bird's-eye view of the marine vegetation are Nahant and Magnolia, the former easily reached by boat or by train from Lynn, the latter by trains on the Gloucester branch. The botanist who is not in the habit of collecting on coasts which have a marked tide should not allow himself to suppose that the beaches in which our coast abounds are good places for obtaining a knowledge of the marine flora at short notice. After the storms of spring and autumn the beaches are often covered with algæ, some of which may be rare; but especially in the summer months they are often quite bare, and it is only on rocky shores that one can always be sure of finding something of interest. The inland botanist should also bear in mind that even on the rocky shores of this coast he will find little or nothing unless the tide be at least half-way out.

Whether the visitor goes to Nahant or Magnolia or other rocky points the species he finds will be the same. The littoral region, or the space between high- and low-water marks, is very characteristic, the rocks and stones being covered with large olive brown rockweeds, or Fuci. At the upper tide-limit is the small *F. distichus* growing in pools, followed lower down by immense masses of *F. vesiculosus* and *F. evanescens* and at dead low tide by *F. edentatus* with long, flat receptacles. *Ascophyllum nodosum*, the largest of our Fuci, with massive, nodose bladders in the stems, is abundant near low-water mark but is not in good fruit in August. Teachers in search of Fucus in fruit should select *F. vesiculosus*, the other Fuci in fruit in this month not having separate male and female conceptacles. In deep tide-pools at low water are to be found the different kelps of which representatives of the groups of *Laminaria saccharina* and *L. digitata* are always to be found, and often with them *Alaria esculenta* in which the fruit is borne in small basal pinnæ. During August the kelps are not seen at their best and usually they do not fruit until autumn. The larger specimens of kelp grow in deep water, but even in mid-summer one may find them washed ashore in sheltered places in the rocks or on the beaches. *Laminaria longicruris*, with long hollow stipes and *Agarum Turneri*, well named the sea-colander from the numerous perforations of the frond, are characteristic of our coast, although the latter reaches perfection only in higher latitudes. The Fuci, as far as the eye can see, form the mass of the littoral vegetation, but they are covered with other epiphytic algæ of which the two most common brown species are *Pylaiella littoralis* and the shorter and denser *Elachista fucicola*. In the tide-pools other filamentous brown algæ abound as well as *Ralfsia verrucosa* which forms irregular, warty crusts on the rocks.

The Florideæ, or red seaweeds, as a rule are more abundant below low-water mark and, like the larger kelps, are to be sought washed ashore in sheltered pools where they have been left by the tide or on beaches. One cannot always count on finding the more beautiful deep-water forms; but one may expect at any time *Delesseria sinuosa*, looking like a red oak leaf, the finely cut *Ptilota serrata*, and *Euthora cristata*. Probably the best place in the world for collecting the beautiful Euthora is the beach at Magnolia after a storm. Certain red seaweeds will be found growing in pools near low-water mark such as *Chondrus crispus*, Irish moss, which covers the rocks at dead low water and is collected for the markets at Hingham and other places, and the digitate, membranous *Rhodymenia palmata*, or dulse, which is sold to some extent in the Boston market. Smaller filamentous species such as *Ceramium rubrum* and several *Polysiphoniæ* also are found in pools. A very abundant red alga is *Polysiphonia fastigiata*, which forms dense, rather rigid tufts on *Ascophyllum nodosum* which do not collapse when left exposed by the tide. The color of this species is, however, almost black, and at first sight it would hardly pass for one of the red seaweeds. The calcareous red seaweeds are represented on our coast by *Corallina officinalis* and several species of Melobesia and Lithothamnion. The Corallina is common in deep pools and is at once recognized by its calcareous structure and pinnate, jointed fronds. Most of the Melobesiæ form small crusts on other algæ and the more solid Lithothamnia form pinkish, stony crusts on rocks and shells, often of considerable extent, and either smooth or rough with solid tubercles.

The green algæ belonging to the Chlorophyceæ reach perfection in mid-summer, whereas most brown and red seaweeds mature at other seasons. The green algæ are most common in the higher pools, even in those where the water is merely brackish or almost fresh. Those of any size may be classed in two groups: the Ulvaceæ or sea lettuces, found everywhere on wharves, exposed flats, and high pools, forming large flat expansions or inflated intestine-like fronds; and the filamentous species belonging to the genera, Cladophora, Chætomorpha and Rhizoclonium. It is hardly possible to specify any of the common forms of this group since the distinctions depend largely on microscopical characters.

The algæ are not the only plants of interest to the visitor to our rocky coast. The maritime saxicolous lichens are very striking to the naked eye but the species are not especially characteristic of the coast. There is one very abundant lichen, however, *Verrucaria mucosa*, which grows on rocks at low tide. It is only necessary to lift up the hanging masses of Fuci to find the rocks beneath covered with the large, dark green patches of the Verrucaria in company with the crustaceous red algæ, Hildenbrandtia and Petrocelis. On the exposed cliffs of the shore will be found fine specimens of the orange colored *Placodium elegans* with other species of the genus, *Lecanora rubina*, *Rinodina oreina* and other crustaceous forms. The branches of the stunted trees and shrubs near the shore are brilliantly colored with quantities of *Theloschistes chrysophthalmus* and *T. parietinus*, common species to be sure, but especially common and beautiful near the sea. The maritime mosses and hepatics are not especially striking on our coast. *Grimmia maritima* is found in small quantities at Nahant as well as *Cephalozia divaricata*, the latter not a common species in eastern New England.

Farlow, W. G. The marine algæ of New England. (Ann. Rep't U. S. Fish Commission, 1879, pp. 1-210. Pls. I-XV.)

www.ingramcontent.com/pod-product-compliance
Lightning Source LLC
Chambersburg PA
CBHW020116170426
43199CB00009B/546